Lean Driven Reliability

Zero to Hero

How to Jumpstart Your Reliability Journey Given Today's Business Challenges

Joe Kuhn

Edited by Kate Marburger

Additional editing by: Lisa, Drew and Kayley Kuhn

Cover design by Valory Waligoski

First Edition: July 2020.

Created by Lean Driven Reliability LLC. All rights reserved.

Liability Disclaimer: This book contains the viewpoint of Lean Driven Reliability LLC based on experiences and opinion. It contains concepts not specific to any one manufacturing plant, and therefore, any action at your plant should be taken with careful consideration to impact on such things as: the environment; health and safety of employees; business results; quality; etc. It is understood that Lean Driven Reliability LLC cannot possibly know all the circumstances at your plant and therefore; results/consequences from actions taken by a company/person from ideas presented in this book are the sole responsibility of that company/person and not Lean Driven Reliability LLC. You should consult a professional for guidance on specific actions to be taken at your plant.

Since this book is the sole property of Lean Driven Reliability LLC, the author, Joe Kuhn, shall be held harmless for all legal matters that may arise from this book.

Contact Lean Driven Reliability at email: leandrivenreliability@gmail.com

Dedication

This book is dedicated to the factory workers. I'm humbled to have worked alongside some great team members. Specifically, I would like to call out some of my professional mentors that went out of their way to help me succeed: Ed Toon, Al Weinzapfel, Royce Haws, Mark Keneipp, Pat Love, and Sherry Shen. I cannot repay the gifts these folks gave me along the way, but I can pay it forward.

I also dedicate this book to my family. My roles as son, husband, and father give me the greatest joy and sense of accomplishment. I have been married for 32 years to my high school sweetheart, Lisa, and have four great adult children: Tyler, Drew, Spencer, and Kayley (all Purdue engineers, by the way). I could not be prouder of them. Lastly, to my mom and dad who gave me a faith-filled, loving, and sturdy foundation; this is the greatest gift any parent can hope to give their children. Thanks Mom and Dad; I'm truly blessed.

Who should read this book?

If you are struggling to start a reliability program, or you're mired in mediocrity with your results, or you've started a reliability program countless times only to "pause" or stop, this book is for you. You will learn new skills that can give you sustainable results, rooted in best practices, beginning in as little as 30 days with little to no investment. Too good to be true? By combining the concepts of lean with reliability best practices, I've uncovered a secret that completely changes the speed of results. This secret is too dramatic not to share. Too many individuals are struggling with a 1980s dogma of how we do maintenance. This struggle is resulting in excessive work hours, frustration, poor results, and a loss of organizational hope. Unfortunately, too many bring this frustration home, impacting family, friends, and communities. There is a better way, and you can start it tomorrow. It has taken me 33 years to learn the lessons contained in this book, and you can have it all in just a few hours of reading.

If you are on the cutting edge of reliability technology and have 20 years of using standardized best practices under your belt, you too will gain new insight to significant waste hidden in your plant. This book could be the boost you're needing to rise to a new level of performance. Expect to be amazed with the simplicity, low cost of implementation, and impact you can have in a very short period of time. Results are sustainable and scalable to any size manufacturing plant. These all drive organizational enthusiasm and long-term sponsorship, which eliminates the starting and stopping most reliability journeys encounter.

Over my 33 years, I have consulted with 32 plants in 8 countries, been a plant manager three times, and worked as a department manager and engineer. 28 of my 33 years were as a practitioner in a plant. Consequently, this book was not written by someone who has studied reliability practices in plants – it's written by someone who had to get it done despite all obstacles. In doing so, I've discovered techniques and lessons not taught at conferences, in trade magazines, in books, nor in requirements for professional certifications. This book outlines the most powerful tools for change I have seen and utilized. You will see new waste opportunities hidden in plain sight with the skills you learn here. This is my promise to you.

Why the Title?

"Zero to Hero" was a phrase I heard all the time from a supervisor of mine, Royce Haws. He always stressed that a relentless focus on basics always wins in the long term despite short term setbacks.

"Zero" for this book refers to:

- You are making *Zero* progress with reliability.
- You have *Zero* plans nor ideas to get better; you're stuck.
- You have *Zero* dollars to spend on reliability.
- There appears to be *Zero* alignment between groups on actions at your plant.
- You have *Zero* hope that next week, next month, or next year will be better.

"Hero" for this book refers to:

- You discover a simple, executable, and sustainable path to a reliability culture – an alternative to your plant's failed attempts at change.
- Your new solution solves the short-term financial challenges that leaders face each month while delivering strategic change.
- You attain organizational alignment, enthusiasm, and hope (from the CEO to the shop floor).
- The *Hero* is not a person, but rather, a new process.
- The team at your site all wins.

How this book is organized

This book is formatted both for your learning today and as a quick reference guide for the years to come. Consequently, I repeat some guidance and lessons learned in more than one chapter of the book. I begin with a reliability deployment story starting on day one of the change process. Stories depict the human interactions along with the tools and processes.

The last chapter is titled "Mistakes." This is a list and discussion of top mistakes I have made along the way or mistakes I have helped others overcome. Knowing these common traps in a reliability journey is of equal value to your skills and best practice tools. Reliability journeys do not fail because of a lack of knowledge; they die because of mistakes. Do not skip this chapter. If you have been on a reliability journey for some time, it is not a bad idea to read this chapter first.

A quick reference guide is in the back of the book. The intent is to provide you with easy access to the concepts of lean driven reliability.

I'd love to hear your feedback; look me up on LinkedIn to reach out.

Let's get started.

Table of Contents

Who should read this book?..4

Why the Title?...6

How this book is organized..7

Preface..10

Chapter 1 Welcome to CoatCo..14

Chapter 2 Detailing the Business Case.............................17

Chapter 3 Finding Current State..20

Chapter 4 Establishing the Target State............................41

Chapter 5 Actions to Achieve Target State......................45

Chapter 6 Business Impact...59

Chapter 7 Measures of Success...65

Chapter 8 Reflection – What have I done?.......................67

Chapter 9 Next Steps..70

Chapter 10 Advice...72

Chapter 11 Selecting a Consultant....................................78

Chapter 12 Selling Reliability..86

Chapter 13 Mistakes..92

Quick Reference Guide..…..…106
    Core Beliefs of Lean Driven Reliability….......107
    Format of an A3.................................…......109
    7 Forms of Waste................................…..……110
    Rules in Use..…...…112
    Chalk Circle Observation.......................….......114
    ADKAR...….......115
    Warren Buffett's 5/25 Rule.....................…..….116
    Listing of Common Current State
    Waste and Potential Actions...................……..117

Preface

Just a little over a year ago, I retired after 32 years with the same Fortune 200 aluminum company. However, the company I began with in 1987 was not the same I retired from in 2019. I don't want to debate over which company was better, but it is critical to note the management philosophies were vastly divergent. In 1987, decisions and investments were always made with a long-term view. For example, one plant I was at bought 100 years of coal reserves. Today, that same plant struggles to secure approvals to purchase coal for the next year. In 1987, investments were made because they made good engineering sense and assumed we were going to be in business in 20, 50, or 100 years. Granted, some of these decisions did not work out, and you could be stuck with the consequences for generations. Nevertheless, the long-term view was the norm. Fast forward to 2020. That same company looks at quarter-over-quarter returns. In good times, we needed to compete for capital with sister plants, and you needed about a two-year full return on investment to get approved – two years! In bad economic times, this was reduced to one year. At the same time, the process to get approval for investment expanded from a few weeks to several months and even years. We did make better decisions, but we were making very few of them and missing dozens of opportunities. Further, and unfortunately, because of the aluminum business cycles what happened all too frequently was the market turned down in the middle of the project, causing it to be delayed or cancelled.

Complicating the situation, top managers are only in their roles three to five years. Thus, they need to provide results in less time to secure annual performance driven bonuses and promotion. At the same time, fewer and fewer leaders have worked themselves up from shop floor positions. To be factual, rather than critical, more and more come from a financial, sales, and marketing background. They may be fantastic with many facets of the business, but – let's face it – not the technical understanding needed to make equipment more reliable. What they do know are numbers. Consequently, many great ideas for the long term fail to get the interest of the leader: "5- to 10-year payback? Get out of my office." They would rather invest in something they know well. Examples include new technology, new assets, increased processing speed, better marketing, cutting cost, and reducing headcount.

What does this have to do with reliability and maintenance? That same divergent gap in thinking applies to all projects, programs, and improvement initiatives. In the 1980s, several leading companies started a reliability journey, knowing it would take great upfront investment, with returns beginning in years four or five. Full return on investment could take 10 years. Many veterans of these companies are writing books and consulting others to follow this path. I have no issue with this approach; it works. However, in 2020, for 95% of plants out there, this sales pitch does not solve the challenges and expectations of the plant manager. They are expected to show quarter-over-quarter improvement. Adopting the traditional deployment would result in termination of the plant manager. Simply put, most reliability professionals are selling something the plant manager just can't buy.

I began my career as an engineer but quickly felt called to supervision (or rather nudged, by Ed Toon, who I mentioned above). Over the years, I advanced to larger and larger leadership assignments. While I learned many critical lessons in each of my endeavors, I want to detail four roles which are critical to the lessons in this book. I was plant manager at an aerospace manufacturer (1000 employees), a power generation plant (150 employees), an aluminum smelting and rolling complex (2000 employees), and the global director for reliability and maintenance (31 locations in 8 countries). Harnessing the hard work, passion, and leadership of others, together we rapidly and sustainably delivered significant reliability results rooted in best practices. Tangible results were visible in weeks with minimal investment. As a general rule, the methods we employed were getting at least 10% reduction in repair and maintenance (R&M) cost in year one, net of any investment (zero investment in most cases). I've seen production increase as much as 29% (still in year one). Over a five-year period, one plant had a 42% reduction in R&M cost and Overall Equipment Effectiveness (OEE) increase from 47% to 84%. In my director role, while facilitating and coaching dozens of others through this process, the plant with the lowest return plan crafted actions to reduce $3MM in R&M spending in 12 months (10%).

I have chosen to share my reliability insights and deployment practices through a story (Chapters 1 through 9). This book tells the narrative of a fictional plant CoatCo. CoatCo is a small plant that boast 100% reactive maintenance, no preventive maintenance (PM), no predictive maintenance (PdM), and no action of any kind to prevent failure. There are no planners, technical assistants, or reliability engineers. The staffing consists of a maintenance manager and 10 craftspersons. I have just been hired as the maintenance manager. The story begins on my first day. The details of CoatCo are a collection of my real experiences from dozens of locations; any similarity to a specific plant is purely coincidental. The circumstances are realistic based on my firsthand experience. In writing this book, I did not survey anyone, primarily because I find surveys entail a heavy bias, deliver overly optimistic or pessimistic information, and omit most obstacles. As the practitioner knows, the devil is in the details, not the highlighted summaries. This is a story that I hope empowers the reader to take action and accelerate their journey out of the muck of mediocrity towards excellence. The concepts and techniques I use are scalable to any site regardless of size and state of reliability (whether you're just starting or you're the industry leader); I've employed these practices at plants as small as 50 employees and as large as 2000. The process does not care; it works equally well in each.

This book is intended to "jumpstart" your journey. From success, you earn creditability; from creditability, you earn future investment. Once you get the flywheel of a reliability culture moving, best practices become easy to justify as only a portion of the dollars already saved. This solves the challenge of the plant manager – quarter-over-quarter results.

I put plants into one of three categories: industry leader; mired in mediocrity; and ignorance. Remember that ignorance does not mean stupidity, but rather the lack of knowledge or information. There are lessons in this story for all types. Industry leaders will learn to focus more on shop floor waste through lean principles as well as their best practices and new technology. Those mired in mediocrity will get a simple, proven formula to accelerate. The plants that don't know what they don't know will learn a process that yields results and enthusiasm leading to greater understanding by all – especially decision-makers.

If you like this book, or even if you don't, please check out my YouTube channel "Reliability Man." This channel is dedicated to giving away my leadership, culture change, and R&M knowledge to the masses. The videos range from 1 to 20 minutes in duration; consequently, for those with little time, this may be an ideal place to learn. Each video has an action you can apply next week for improved results. I have three goals for my channel:

1. To positively impact those individuals that can't afford a consultant yet need exposure to reliability best practice implementation challenges and solutions.
2. To use a popular media platform to advance the next generation of leaders in creating a reliability culture.
3. To offer an alternative to traditional reliability deployments that solves the business challenges of today.

My millennial children tell me they will listen to my YouTube videos long before tackling a book. My stated mission in the content of each video is to "bridge the gap between best practices and the reality you live in every day at your plant." I have over 80 videos posted as I write this book.

Chapter 1
Welcome to Coatco

CoatCo has 90 employees. It is considered a heavy manufacturing facility – that is, it employs large equipment to make its products. The plant cleans, levels, coats, and lubricates bare steel and aluminum sheet coils for the food and beverage industry. The bare coils are supplied by rolling mills located in the Southern and Midwestern United States. After the coils are coated, they are shipped to a can maker, where the sheet is converted into cans. A typical coil is 60 inches wide, 0.007 inches thick, and 45,000 feet long. Final products at the can maker include soft drink cans, beer cans, and pet food cans. Coating is required to protect the contents of the can from damaging the metal can. Lube is required for the customer's can press process.

CoatCo was built in 1985 and consist of two coating lines and support equipment. The first coating line was installed in 1985, and the second in 1992. Unfortunately, the coating lines are from two different manufacturers, and let's say have their own personality and product preferences due to design differences. The plant operates five days a week/three shifts a day, with some limited overtime on weekends. CoatCo only operates 50 weeks per year to have a two-week shutdown in December. The coating process is continuous. Each coil is unwound, leveled, cleaned, coated, cured, lubed, and then rewound. The head (or beginning) of the next coil to be produced is mechanically stitched to the tail (end) of the previous coil. This joint is made while the line is running due to an accumulator that rises while running the main body of the coil, then lowers to permit time for the joint to be made with a punch die. Typical equipment on site includes rollers, motors, coating heads, ovens, burners, pumps, sprays, filtering, a tensioning system, gearboxes, and bearings. Apart from the production centers, there are many systems that support the coating lines. CoatCo calls these processes "off-line equipment." A sampling of off-line equipment on site includes air compressors, cooling towers, waste water treatment, coil unloading equipment, a coil packing line, several automated doors, HVAC systems, lighting, power distribution systems, overhead cranes, and four fork trucks. Many of these systems are required to be operational for the lines to produce coils; however, some can be down due to buffer capacity (storage tanks, for example) or having inline spares.

The maintenance staff consist of 10 craftspersons who are all multi-craft skilled (trained to do both mechanical and electrical work). Four are scheduled to work day shift, three on afternoon shift, and three on midnight shift. Plant historical experience concludes it takes three persons to perform reactive maintenance on each shift. The fourth employee on day shift provides vacation coverage for the other nine, manages the tool and parts crib, and expedites parts. Outside contractors are hired to do all rebuilds and major maintenance, such as widening the line, speed improvements, or coating head design changes. Rounding out the maintenance team is a maintenance manager – my new role. I report directly to Jane, the plant manager. The previous maintenance manager left the company two months ago after giving his two weeks' notice. It is important to note the plant has zero maintenance planners, zero technicians, and zero maintenance and reliability engineers. No key performance indicators (KPIs) are tracked. There is no monitoring of spending nor a department budget. The current state equipment maintenance plan is 100% reactive maintenance. That is, when it's broke, we fix it. Production operators call craftspersons via a radio to alert them of equipment failures or concerns.

The plant leadership team consist of the plant manager (Jane); production manager (Jeff); maintenance manager; sales manager; production planning manager; the environmental, health and safety manager; quality and improvement manager; human resources manager (Jordan); and a controller (Jim). The plant manager reports to the president of North American Operations. CoatCo has six other sister plants that coat sheet for other industries. The president reports to the Chief Executive Officer of Precision Manufacturing, Inc.

It's 7 a.m. on Monday, and my first day on the job. After some brief orientation, I have a lead team meeting at 9 a.m. The meeting was set up weeks before my arrival so lead team members could prepare to share the business case for change within plant maintenance and reliability. After brief introductions, Jane turned the meeting over to me to facilitate. I began by introducing the team to the A3 process. The A3 process is a structured procedure for problem solving with origins in continuous improvement and lean. Toyota was the first to utilize the tool. The process follows a rigid flow that guides the user through a logical problem-solving sequence. The name "A3" comes from the size of paper originally used when A3s were hand written. Presently, most practitioners document the process electronically, but the name A3 survives. An A3 has five sections: Business Case, Current State, Target State, Actions, and Measures. Today, I explained, in the next two hours, we are just going to complete the Business Case. The Current State will be derived from shop floor observations. The Target State will be completed in the conference room based on the Business Case and Current State. Actions will be crafted to achieve the Target State, and Measures refers to how we agree to monitor our progress towards the Target State. The whole A3 process will take place over the next two and a half weeks. Current State observations will make up the lion's share of our calendar time and is the most critical. Any questions on process?

Chapter 2
Detailing the Business Case

Still in the lead team meeting: We are here to collectively define and understand the opportunity of improved reliability and maintenance performance to bottom line business results. You have been asked to bring both data and opinions. Are we trying to lower cost? increase capacity? improve quality, safety or environmental excursions? After we define the opportunity, we will answer, "If we do an excellent job starting today, where could we be in 12 months?"

After about an hour of brainstorming with yellow Post-it notes, we combined and eliminated some opportunities before arriving at the following five items in rank order:

1. Increase Capacity: We can sell 15% more production – combined, unplanned maintenance downtime for both lines is currently at 35 hours per week. (120 calendar hours times 2 lines equals 240 total hours. This calculates to 14.6% unplanned downtime for maintenance.) One hour of downtime is $500, or $17,500 per week, or $875,000 per year. From competitor research, we have estimated best-in-class to be 5% unplanned downtime for equipment issues on similar coating lines.

   Note: The plant manager and president really do not want to go to seven-day crewing to bolster capacity. This business can cycle, and they don't want to go through the hire/layoff game. A family culture is a value they plan to protect. Both believe an improvement in reliability is the best solution.

2. Scrap Generation: Since our processes are continuous, every downtime event results in 2,000 feet of coated scrap at $0.25 per foot of lost revenue and re-melt cost (re-melt involves shipping the coated scrap to a contractor and having them melt the scrap

down into the base metal while burning off the coating). In an average week we have 21 unplanned maintenance events, or $10,500 per week, or $525,000 per year.

3. Customer Quality Returns: Rejected metal from customers averages $100,000 per month. Our root cause problem solving has determined that 20%, or $20,000 per month, or $240,000 per year, goes toward equipment issues related to reliability.

4. Maintenance Costs: We do want to better understand our costs and put controls in place, but this is a much lower priority than items 1-3 above.

5. Safety and Environmental: We currently are performing quite well in these areas by all measures but believe our systems can be more robust. Again, this is a lower priority than items 1-3 above.

Total Opportunity: $875,000 + $525,000 + $240,000
= $1,640,000 per year

Lastly, as a team we defined "Excellent in 12 Months" as having a run rate that is 20% lower than current performance, with systems in place to drive results to a 50% cut in waste in 24 months. A 20% cut would mean $328,000 saved for the business.

This completes the Business Case.

As a learning opportunity for the lead team, I asked the team to brainstorm actions they think we need to take based on their knowledge of manufacturing and this new Business Case. Four actions surfaced:

1. Hire an engineering firm to detail preventative maintenance tasks for our assets.

2. Hire three to five more maintenance personnel and begin weekend overtime to complete the preventative maintenance tasks.
3. Hire an outside firm to recommend and create predictive maintenance routes and to perform these routes. Anomalies (defects) found on routes will be communicated to our maintenance resources for repairs.
4. Get some discipline in cost controls – specifically on buying materials. This is where we tend to have surprises in the monthly costs. Consider having the plant manager approve all spending above $100.

We will save these actions for a later discussion.

Chapter 3
Finding Current State

Background

Understanding the Business Case is critical to calibrating your senses to look for systemic waste and inefficiencies. The lead team meeting was a great start. Next, we need to understand reality on the factory floor. I have a six-part process for finding Current State:

1. <u>Interview my peers one-on-one on the shop floor</u>: I set up two hours with each member of the plant leadership team (a total of seven peers) to walk and talk out in the factory. Candor level dramatically increases in this environment. Statements will be made that confirm, supplement, enhance, or even outright contradict the group lead team meeting findings. This is critical in your pursuit of reality through observation. The lead team discussions and the peer one-on-one's direct where to observe in the plant. This saves time. In some plants, key performance indicators (KPIs) are also used to focus the observer on where waste may exist.

2. <u>Chalk circle observations</u>: Taiichi Ohno of Toyota fame defined the term "Chalk Circle" observation. Chalk circle refers to an imaginary circle drawn on the factory floor in which the observer is confined for at least four hours. Eight hours and multi-day observations are preferred as well as employing multiple observers (the number of observers depends on plant size). Many leaders observe for two minutes and draw conclusions based on this snapshot of reality. The premise of chalk circle is that you really cannot grasp the intricacies of a process in just minutes. While doing chalk circle, it is common for me to think I'm wasting

my time, but then after two hours – bam! There it is – a gold nugget of waste. The vast majority of leaders spend no time in chalk circle and thus are doomed to make bad decisions based on incorrect data. The irony is that if you interview leaders, they will tell you they are a "floor leader." This is the disease impacting manufacturing today – leaders are making million-dollar decisions based on biased opinion, filtered electronic data, and an illusion that they have observed reality.

3. <u>Draft common themes / opportunities</u>: From my observation notes, Business Case, and peer interviews, I group together common themes and prioritize. I typically try to get to four to six high impact common themes. More is not necessarily better; our goal is to identify major leverage points that will impact the business. Acting on 20 leverage points will greatly dilute our attention and thus lower our success rate for implementing change. Focus is key. These common themes are NOT solutions. They are cultural norms of current state. Solutions will come later; for now, let's work the A3.

4. <u>Review common themes with stakeholders</u>: These draft leverage points next need to be reviewed by those involved in the observations. For example, if you observed a team of two mechanics, review your findings with them. Repeat until all parties involved have had the opportunity to see your leverage points. Caution: Be prepared for comments like, "This never happens." Or, "This was a very unusual day." These should be trigger words for you. What are the odds that the waste you observed was the first time it has occurred in the plant's 25-year history? It is highly likely the waste you observed is common, and employees have been conditioned to ignore it.

5. <u>Make any modifications to common themes / opportunities</u>: If you receive consensus on your common themes, they are

ready to present to the lead team. However, most likely one or two of these themes will be challenged, or you will be told they are wrong. In this case, you must not discard or alter the theme, but rather perform more chalk circle observation. Observation trumps opinion. If the disagreement persists, you need to hold fast to your leverage point. You may end up being wrong, but more likely you found a diamond in the rough.

6. <u>Present current state to the leadership team for acceptance</u>: Gather the lead team together to review the depth of your observation process, common themes, and review process. You do not need to share all observations, but ensure the team understands the number of people, processes, and hours you invested. As stated before, observation will trump opinion. For example, stating that we lubricate each roll in the accumulation tower (50 rolls) once a week is vastly different than stating, "I have observed the last four lube PMs on the accumulation tower. In three of the four PMs, the mechanic missed seven, nine, and ten rollers. In the fourth PM, zero rollers were lubricated due to time constraints." Which statement do you believe? The latter wins every time. That's the power of observation.

You need to push for total acceptance of common themes from the lead team. If you do not, you need to observe more and repeat the process. Invite those that contest your conclusions to join you in chalk circle observations. I have never had to do this.

I have been highly trained on two more tools critical to seeing the real current state. They are:

1. Seven forms of waste
2. Rules in use

Knowing the seven forms of waste is fundamental to lean manufacturing. The wastes are Transportation, Inventory, Motion, Waiting, Over-Production, Over-Processing, and Defects. (Hint: A simple acronym can help you remember these: TIM WOOD.) I will give brief descriptions of each, and you can find much more on the internet.

Transportation Waste – Moving product, materials, tools, or equipment from one place to another on the shop floor. In many cases, this movement is required, but ask yourself if the customer is willing to pay for it. Regardless, it needs to be reduced as much as possible. Example: A mechanic walking back to the shop to get a wrench. Yes, the mechanic needed the wrench, but why did she have to make a second trip?

Inventory Waste – Having money tied up in excess inventory. Example: Having five spare motors in stores when your annual consumption is one.

Motion Waste – Non-value added movement by employees. Example: A mechanic installed the wrong pump. Upon realizing the mistake, he found the correct pump and installed it. The error extended the job 3 hours.

Waiting Waste – Waiting for instruction, parts, processing, or help from others. Example: Two mechanics waiting for a part to be delivered from the storeroom.

Over-Production Waste – Making too many parts. Example: Maintenance built 10 molten metal pump systems as spares when monthly consumption is one.

Over-Processing Waste – Performing more work on a process or material than is necessary to meet a standard. Example: A plant has a lube sampling program as part of its predictive maintenance plan. If the oil is changed monthly on a PM, the PM would be considered over-processing. It is a step that is not required to meet a standard.

Defects Waste – Non-conforming material – material does not meet the standard. Example: A new motor was installed on the crane. The motor shaft was not aligned properly, and the newly installed motor failed in two days.

> It is critical for you to understand that EVERY reliability tool and best practice is attempting to *eliminate waste*.

Rules-in-Use: In a paper titled "Decoding the DNA of the Toyota Production System," Steven Spear and H. Kent Bowen attempted to break down the culture that exists in a lean plant. They simplified lean to four rules:

1. The Work of One
2. Connections
3. Fixed Flow
4. Improvement

I will further simplify and apply the concepts to reliability and maintenance.

The Work of One – The tasks of each person working are highly specified for content, sequence of steps, time required, and outcome. Example: If a mechanic is to replace a fan belt, does he know all the steps to perform in sequence within a specific time window? Lastly, how is the work tested to ensure the repairs do not contain defects? A good job plan provides this information.

Connections – Rules or norms exist between workers on how they work together or how work is transferred from one to another. Example: After the electrician locks, tags, and verifies isolation of the overhead crane, he radios the mechanic to begin his work (after he puts his lock on the lock box). Connections have proven to be a major source of waste in my experience – most especially connections across shift changes.

Fixed Flow – There is one right way to do work. Example: Work orders are submitted electronically to the planner by the requester. The planner does not accept phone calls, meeting comments, nor hand-written notes as work orders.

Improvement – There is a system to drive improvements, and everyone knows it. Example: We have biweekly OEE meetings for our production center.

*****

Now back to CoatCo.

I had my meetings with six of my seven peers with little new information from the lead team meeting. It was great to get to know them better, and all were welcoming and eager for me to help the plant perform better. I did have an interesting meeting with the production manager that is key to the A3 process. Here are my notes from the conversation:

> The previous maintenance manager left the company in frustration with Jane, the plant manager. Jane did not support his plan to turn things around.
>
> Previous maintenance manager plan:

- Hire a planner. Full up cost $120,000 per year and cost of formal planner training
- Hire a reliability engineer. Full up cost $120,000 per year and cost of reliability training.
- Hire 3 more craftspersons to execute planned work. Planned work will occur on weekends on overtime. Full up cost $120,000 per year times 3 persons, or $360,000 per year.
- Hire an outside contractor / PdM firm to perform infrared, vibration, and lube system sampling. Estimated at $160,000 per year.
- Lease a Computerized Maintenance Management System (CMMS). $12,000 per year, plus $50,000 for initial setup.
- Expected increase in materials, parts, and labor (weekend overtime) to get equipment in maintainable shape. $50,000

per month for 18 months. For year one, this equals $600,000. For year 2: $40,000 per month, or $480,000 per year.
- Total year one expenses: $1,400,000 ($1.4MM) with year 2 forward estimated at $1,250,000 ($1.3MM).

Recall that 100% of the opportunity from the Business Case is $1.6MM. There is no realistic possibility of getting all the savings; 70% was proposed as the target for year five by the previous maintenance manager: 70% of $1.6MM is $1.1MM. Consequently, the proposed plan was to invest $1.4MM, $1.3MM, $1.3MM, $1.3MM, and $1.3MM over five years (total = $6.6MM invested). Savings from the newfound reliability would be 10% in year one, 20% year two, 30% in year three, 50% in year four, and 70% in year five (total savings from improved reliability = $2.9MM). After year five, the investment and savings are roughly even.

In summary – huge upfront cost with really no chance of returning investment (ignoring the time value of money). Now, to be fair, there is opportunity cost associated with poor reliability. Staff resources will be able to spend their time on innovation, design changes, speed improvements, and quality differentiators as well as other proactive actions to drive results. Nevertheless, this is a leap of faith the plant manager refused to endorse. Recall the tenure of most plant managers is three to five years. Further, recall the plant manager is expected to have quarter-over-quarter improvement. Lastly, the plant manager knows she will not be able to maintain full sponsorship for five years due to business cycles. This plan was dead on arrival.

> The production manager gave the following advice: "Get efficient with what you have first; get results, then ask for investment from a position of creditability."

I was very glad I spoke with the production manager. Not only will I need him to succeed, but also, he provided valuable insight to how the plant manager thinks and acts.

Chalk Circle Observations

This section is the most critical of the entire A3 process and of this book. Why? Because it is the step most often skipped or rushed through. A mentor of mine told me to "stay in current state three times as long as I felt necessary." In other words, if I thought four hours of observation was sufficient, I needed to spend 12 hours observing. From my 33 years of industry experience, I contend that several billion dollars have been wasted in manufacturing by "jumping to conclusions" from the business case. Smart managers know from experience what works and what does not, right? I'm sure everyone has war stories they can insert here. My favorite example was in a process called scalping in the manufacturing of aluminum sheet. The scalper was frequently the bottleneck of the plant. Over the years, we spent $250,000 on equipment upgrades to improve output; all were unsuccessful. A new leader came to that department with a lean background. He began with observation. He discovered significant waste in the connection between shifts and the sequencing of employee breaks. He implemented no-cost procedural changes that yielded a 22% improvement to throughput in under 30 days. This improvement has been sustained and improved upon for the last 20 years.

My first step in observation is deciding what to observe. As stated before, this is where the business case and peer observations are critical. In plants with key performance indicators (KPIs), the A3 leader should add these to the data set. Recall CoatCo has no KPI data. I chose to observe the coating line process, the coating line operators, and all three shifts of maintenance crews. The coating line process and operators, I felt, could be observed together. My goal with this observation was to really understand how the equipment was being used. In my experience, at least 50% of opportunities for reliability improvement lie with operations. Think of your car. Your driving habits and attention to preventative maintenance schedules has as much to do with the reliability of your car as does the shop mechanic; actually, your impact is much greater. This does not mean trying to place blame, but rather being open to where the waste may present itself. My first step is to get sponsorship from my production manager peer to observe his people and processes. I need to assure him that we are in this together, and I would present all findings to him prior to exposing the lead team to the information. Fear of embarrassment is an important emotion to manage with chalk circle observation.

Recall from the business case and peer review that we are experiencing a large number of unplanned downtime events for equipment, the duration of events is unpredictable, and we have no process to improve (we have 100% reactive maintenance with no problem-solving system). Without a design to improve, there's a good chance that next week, next month, and next year will look a lot like last week, last month, and last year. My favorite question to ask is, "Why will we be better next month?" This information, along with our seven forms of waste and Rules-in-Use tools, will enable us to see waste, which we will turn into opportunity.

When observing at a plant where you expect to see a lot of waste (such as CoatCo), do not document five seconds of waste. We are looking for waste no smaller than five minutes. This may need to be adjusted based on the volume of waste observed. For example: If you have 50 five-minute or more observations of waste in a day, you may want to move your threshold to 10 minutes to only catch the big wastes. Conversely, if you have two events documented in four hours, you may lower your trigger to three minutes. Remember, we want to end this step with just four to six common themes. Tools needed to observe include a note book and a pen or pencil. I do not recommend a stopwatch; employees find them intimidating. I use my watch to record time. It may be a good time to reflect on all steps of the A3 at this time so you keep in mind where we are going next. Specifically, do not jump into actions or solutions at this point. Your goal is to document current state reality.

The first five minutes of the observation process is critical. This is where you meet the participants in your observation process. The first minute can be quite defensive and what I call "crusty." There's a good chance the employee will think, "This bozo management person is going to stand over my shoulder and point out everything I do wrong." Also, a good chance the employee will have a lot of withheld management feedback bottled up from years of experiences with leaders. But just trust me – I've done this hundreds of times. It takes about one minute to "break through the crust" (often highly justified) to reach the real person underneath that really wants to do a good job every day and be part of a successful team – one minute of awkward conversation interspersed with painful silence. You can do this.

Here is what I said in my greeting.

I'm the new maintenance manager. I'm here to fix things that frustrate you. The equipment is not running right, there are too many temporary patches of duct tape and chicken wire, you don't have the right tools or parts, we don't PM the equipment, and nobody listens. I understand. I have some ideas and experience with best practices I think can be of value here. What I want to do for the next two days is to walk a mile in your shoes. I don't want you to do anything different. I will want to tap into some of your ideas. I will be looking at connections, such as between you and maintenance. I will be looking at the system in which we expect you to get your job done, and how you set priorities.

This is not an evaluation of your work skills, mistakes, nor work ethic. I have conservatively observed over 500 persons and have never concluded the worker lacked skills or made mistakes – never. It is always the system in which they are forced to work that holds 99.9% of the waste and is the target of my time.

Here is my custom crafted observation schedule and process for CoatCo. I keep this in front of me at all times to keep me on track. It is very easy to get distracted or drift off task while observing; the biggest trap is to jump to solutions to fix waste (especially for engineers). Keep your focus here on observed waste.

A. Coating Line 1: Wednesday/Thursday - 18 hours over 2 days
   1. Spend 3 hours on line entry, then 3 hours on line exit – repeat.
   2. Ensure I see 2 shift changes per day (I'm looking for connections waste).
   3. Look for the 7 forms of waste.
   4. Look for Rules-in-Use: Standard work for one, connections, fixed flow, and improvement. Do I see any evidence of problem solving?
   5. At the end of each day, record wastes: Document a hypothesis or two, group together common themes, and prioritize common themes based on estimated business impact. Use "ballpark" estimates like $10,000 or $100,000 or $500,000. No need for calculations at this point.
   6. Critical: Do not discount waste observed because someone else tells me that "this never happens." Observation is for facts, not opinion.

B. Production Line 2: Friday/Monday - Repeat 18 hours of observation.

This completes the production process observation. Fun, right? It is very tempting to think you're wasting your time one, two, or even three hours into an observation. Stick to the plan. I have never left a chalk circle observation without profound new knowledge of how the business operates – never. Next step: the schedule for the maintenance team. The total maintenance observation time is 6 days. I know this seems long, but I'm following the rule of spending three times as long as you believe to be adequate in observation.

C. Observe 1 maintenance person on day shift for 2 days.
   1. Use the greeting above to break through the crust.
   2. Ensure I capture 2 shift changes per day (looking for connections waste). Start of shift and end of shift usually have a spike in waste.
   3. Look for 7 forms of waste.

4. Look for Rules-in-Use, standard work for one, connections, fixed flow, and improvement. Do I see any evidence of problem solving?
5. At the end of each day, record wastes: Document a hypothesis, group together common themes, and prioritize common themes based on estimated business impact.
6. Critical: Do not discount waste observed because someone else tells me that "this never happens."

D. Observe 1 maintenance person on afternoon shift for 2 days.
E. Observe 1 maintenance person on midnight shift for 2 days.

Next, I analyze all my observation data from operations and maintenance, looking for common themes and repeated waste. I also need to estimate business impact. Sometimes this is a calculation (see below), and sometimes a gut feel estimate.

> Example: I observed 20 shift changes with my plan (8 with production and 12 with maintenance). On 10 shift changes, one of the lines was down for an equipment issue. On average, shift change added 62 minutes to the downtime. So the math looks like:
>
> **62 minutes ÷ 60 minutes in an hour x $500 per hour = $517 per event**
>
> I observed 10 events out of 20 with extended downtime, or 50% of the time. We run 50 weeks a year, 5 days a week, with 3 shift changes per day:
>
> **$517 x 50 x 5 x 3 x 50% = $193,875 per year impact**
>
> I would round this to $200K. This will be used in current state below.

Next, prioritize your ideas based on impact. Keep in mind, you are looking for actions you can implement in under 90 days with very limited investment. Focusing on the 7 forms of waste and Rules-in-Use should guide you to such conclusions. Merge your observations with insights you gained from the business case and peer discussions. Match observations to the most applicable business challenge.

Again, give more creditability to your observation data than your KPIs and documented judgements by others. Why? Observations by a trained observer are less biased, and therefore, are a better reflection of reality. Secondly, observed waste tends to yield solutions that are simple, with very low to no cost.

During my 10 days of observation at CoatCo, I filled up a notebook with hundreds of waste observations. Per the process, at the end of each day I reflected on major observations, common themes, and impact. I picked the big 4 to 6 insights per day. I then compared these insights with the business case and peer review to finally get to the list below. These are my raw observations after:

- Grouping together common themes
- Discounting those of low business impact
- Removing those that took excessive capital investment, and
- Removing those that would take longer than 90 days to execute.

You might expect to have a list much longer (10 days times a minimum of 4 levers per day equals 40 total leverage points). However, there were many like observations that were made over multiple days. This is great news; this means these are not one-off events. Presently, the list below is still in somewhat random order. Here are my notes.

1. Craftsmen divide up equipment into 3 areas: Line 1, Line 2, and off-line (compressors, fume exhaust, water treatment, coating mixing, warehouse, HVAC, etc.) Each craft takes calls related to the assigned area only. It was common to observe one mechanic busy with 2 to 4 breakdown calls while other crafts were at zero calls. Since I was following just one mechanic, I do not have an actual count; however, it was at least 8 of 10 observed shifts.

(Estimated impact: $100K)

2. Communication across shift change is poor and not standard (for both production and maintenance). (Estimated impact: $200K; calculation provided previously)
   - Events across shift change take a minimum of 62 minutes longer. Observed 20 shift changes – observed 10 events with an extra 62 minutes.

3. 10 of 10 shifts (100%) observed had at least 2 hours of consecutive time with no emergency calls (no calls for any of the crafts on shift). Worse, 4 of 10 shifts (40%) had 4 hours of consecutive no-call for reactive maintenance. (Estimated impact: $200K)

4. Craftsmen decide on their own to work overtime – whether they come in early 2 hours and/or leave 2 hours late. 6 of 6 shifts observed had 2 hours of overtime by all craftsmen.

   - 2 hours of overtime per day per craft:
   10 crafts x 2 hours/day x 5 days/week x 50 weeks = 5000 hours per year
   - At $30/hour base pay, overtime is paid at 1.5 times base, or $45/hour:
   5000 hours per year x $45/hour= $225,000/year estimated impact

5. Wrench time (WT) for the 6 days with crafts was calculated to be 12%. (Estimated impact: $300K (one extra person per shift required due to low wrench times))
   - Main detractors: No work, procurement or lack of parts, waiting.

6. Several emergency calls were made for equipment not on the coating lines and not requiring repair within the next 48 hours. Example: Broken roll up door. I observed an average of 5 events/shift, or 8 total hours of work. In essence, one of

the three craftspersons was doing off-line work not critical to line operation full-time (all this at 12% wrench time). (Estimated impact: $100K)

7. Several emergency calls were repeats of previous work (either this week or last). Crafts estimate 50% of calls are repeats; I observed 30%. Examples:
    - Locked up same rollers – 5 times
    - Unwind reset – 7 times
    - Out-of-tolerance tracking of rewind (how straight the wall is on the wound coil) – 8 times

   (Estimated impact: $500K)

8. One craft, John, spent at least 2 hours of every shift working on a line speed increase project. He has been working on it for 3 years. Lacks funding, but John loves the challenge. Is this project sponsored? (Estimated impact: Included in low WT estimate)

9. Another craftsperson, Lisa, lubed accessible bearings at every opportunity. No documentation. She said she was just doing what she can (to help with reliability). She pumped grease into bearings until it could be seen coming out of the bearing. She used the grease that was on the shelf. She had no specifications for how much or which grease to use. (Estimated impact: Unknown, but a serious lubrication/precision maintenance issue)

10. The maintenance team:
    - 10-22 years of experience
    - All good troubleshooters who can restore production flow quickly given part availability
    - Know the equipment well and generally have good skills – lubrication was an exception. Need precision training.

- All appear eager to improve, but don't know how in this culture

11. Operators are not involved in maintenance of equipment – no Total Productive Maintenance (TPM), 5S, clean to inspect, no OEE teams. (Estimated Impact: $300K)

12. Equipment – Common to see oil leaks, poor labeling practices, trash, and chaotic tool storage. (Estimated impact in TPM estimate.)

13. Operators are required to "manage" several equipment malfunctions that have temporary fixes in place. This takes energy and attention – waste. Example: Manual overrides versus automatic sequencing on unwind. This workaround has been in place six months. (Estimated Impact: $200K)

14. Operators do not greet crafts at problem equipment. There's only radio call communication between them – operators go to breakroom, bathroom, or outside to smoke. No standing quality test of repairs upon returning to operations. This adds time to every repair. (Estimated impact $400K)

15. Operations averages 2 coating changes/shift/line. Each change takes 1 hour. Can this time be used for PMs?

16. "Us and them" mentality exists between Maintenance and Production.

17. No shift communication meeting between maintenance and production.
(Estimated impact $200K)

18. No shift communication meeting of crafts to know priorities and help each other as required. "Every man for himself" on his assets. (Estimated impact: $100K)

19. No PM, PdM, or problem-solving work observed. (Estimated impact: $500K)

End of Observations.

Note the sum of the savings is $3.4MM from these estimates, and the total opportunity detailed in the business case was $1.64MM. Don't panic. The business case did not include expenses; it only highlighted opportunity from improved reliability. Further, the $1.64MM is most likely understated based on what we observed, and the $3.4MM is most likely overstated with potential overlap between observations – that is, double counting. The real need at this step is to assign relative value. Differentiate between a $500K change and a $100K change. These numbers will help prioritize our actions. We will refine our savings estimates when we detail the actions we take later in the A3.

Next, I further combined these 19 observations to just five top leverage points. This is my draft of Current State. Again, I target four to six leverage points based on business impact, simplicity of solution, cost to implement, and timeline to implement. Normally, I find the process of going from many key observations to just a few items for current state quite easy and obvious. If you end up with seven to ten, just progress with the larger list. It will get cut down later in the A3 process. If you have more than 10, you need to trim the list now.

Current State leverage points

1. No standards for communication:
    a. Maintenance to maintenance across shift change
    b. Maintenance to maintenance on same shift
    c. Maintenance to production across shift change
    d. Maintenance person to production operator during downtime

2. No work assignment priority; no work leveling between crafts and no overtime approval process.
3. No planned maintenance work to improve reliability: No PMs, PdM, or problem-solving actions. No planned work results in very low wrench times / efficiency.
4. No reliability ownership actions by operations personnel.
5. Crews want to improve, but don't have a system/culture to support – they feel powerless.

The next step I took was to review this draft with the maintenance and production operators that were involved in my observations. This serves several purposes:

1. It demonstrates that you value their opinion,
2. It demonstrates that you are taking the waste seriously by following up with them,
3. It gives them a sense of ownership of the upcoming changes, and
4. It gives you opportunity to find errors or refinements to the current state.

Once again, expect some feedback along the lines of "this never happens" or "you just caught us on a bad day." If their feedback causes you concern, your action is to increase your sample size by observing more. I rarely have to do this. Generally, if I get pushback, I state, "I would like to run a 90-day experiment to see if we can improve. If I'm wrong, I'll buy you a candy bar – deal?" Most of the time, the feedback includes the crafts providing more examples supporting my Current State.

The drafted Current State is now ready to review with my production peer; remember my promise? I reviewed with Jeff. We decided to present the Current State together to the lead team. The lead team meeting on Current State was set up the next day for an hour. I reviewed my observation process, and the 19 top observations. Jeff communicated the five leverage points derived from these observations. I then asked for questions. There were some basic process questions, but as expected, no argument with the observations. Opinion just does not beat observation. I quickly described where we were on the A3 and that defining the Target State is next.

Jane, being a good leader, asked when the A3 will be completed. I said, "The Target State, Actions, and Measures should be completed in draft form in the next two days. Jeff and I will then bring the whole A3 back to this team for approval. If no revisions are needed, we should be complete in three days." Jane loved the response.

The controller, Jim, then asked, "Will these actions meet the Business Case?"

I replied, "That is what the Target State will tell us. Hold the question for our review in three days." Jim was skeptical, but accepted my response. He will only be satisfied upon seeing the numbers – you've got to love accountants.

Points to Ponder

Remember the potential solutions we detailed just after the Business Case? Let's look at them in the light of the new data from observation.

1. Hire an engineering firm to detail preventative maintenance tasks for our assets.

    Comment: Our current maintenance staffing is very underutilized, and they know the equipment very well. If we can free up some hours, we will get far better PMs by using our crafts. It does not seem to be a good cost to incur at this time. Perhaps at a later date, when we get some of the basics in place ourselves. Also, we need to modify the culture to accept the changes that planned work will require. If not, there is a good chance the PMs will just sit on a shelf in my office.

Lean Driven Reliability

2. Hire three to five more maintenance personnel and begin weekend overtime to complete the preventative maintenance tasks.

   Comment: Perhaps working overtime on the weekend is part of the solution, but hiring more maintenance resources when we are at 12% wrench time is poor management – plain and simple. Further, reactive maintenance work has a tendency to expand to the number of people assigned to it. Without a culture of improvement established, there is great risk these new people will just be absorbed with little benefit. Lastly, when the business cycles down, one of the first things we will cut is weekend maintenance overtime, and thus all planned work. This does not appear to be a sustainable solution. I believe this action will be completely ineffective and very costly.

3. Hire an outside firm to recommend and create predictive maintenance routes and to perform the routes. Anomalies (defects) found on routes will be communicated to our maintenance resources for repairs.

   Comment: Our staffing has lots of idle time. Perhaps we can free up a resource and train to complete this work. Our maintenance team knows the assets better, which will enable far superior problem solving. From a culture standpoint, results from PdM will fall into our reactive maintenance culture – most likely being discarded for more urgent reactive work. Lastly, when the business cycles down, there is a good chance we will cut this expense.

4. Get some discipline in cost controls – specifically on buying materials. This is where we tend to have surprises in the monthly costs. Consider having the plant manager approve all spending above $100.

   Comment: I believe we have an opportunity with the lack of a process for overtime approval. Material spending will not improve in a reactive environment, since we only fix what is broken. We can report more but will be powerless to take actions. Efforts to improve planned work should be prioritized over administrative restrictions. We should revisit this action in 12 months.

Clearly, our proposed solutions created with just the Business Case were wrong. We would have spent a lot of money (perhaps $1MM) and have nothing to show for it. We would lose both creditability and time. Further, Jane already shot down a similar proposal from the previous maintenance manager. These solutions also completely missed four of the five common themes we found from observations. The only one that was partially addressed was to increase planned work – but as described, it has little chance of success and would have come at great cost.

Our current state themes point towards very low cost and impactful, quick cultural wins – wins that can earn us some creditability with the plant manager. This creditability can then be used to seek further results that may have financial investment requirements.

Chapter 4
Establishing the Target State

The Target State must be created from the five Current State leverage points. For this step, I like to facilitate a brainstorming session with the lead team. The actions that come out of this step will need to be actively sponsored and audited. I will need full lead team participation to make this manageable; consequently, this needs to evolve into the lead team's plan, not Joe's plan. The format for the meeting:

Meeting Goals
1. Create meaningful organizational momentum towards a reliability culture to create excitement. Everyone likes to be part of a success.
2. Actions must be simple, executable, and sustainable; rooted in R&M best practices; and lay the foundation for future improvement. No short-term only wins.
3. The Target State must be deemed achievable by all CoatCo employees

Agenda
1. Current State leverage points will be emailed to all lead team members two days prior to meeting.
2. We will brainstorm a Target State condition based on improving the leverage points from Current State.
   The team will align on four to six Target State conditions we wish to create within 90 days. We want as many actions as possible in less than 30 days.

Meeting Logistics / Rules

1. The meeting will be four hours at an off-site conference room to aid in creativity and focus.
2. If actions cannot be agreed upon, consider selling it as a 90-day experiment – why? The word "experiment" implies we intend to learn and adapt our actions from the new insights. An experiment will only become permanent if successful. I've sold a lot of changes with this qualifier.

We held the meeting with all in attendance due to excellent sponsorship of our plant manager, Jane. Our final Target State, after just 60 minutes of discussion, is shown below. No doubt involving the lead team from the beginning saved hours here.

Target State

1. Standard work (written procedures) exists for communication between these parties listed below. All shift-to-shift communication shall include: EHS (Environmental, Health and Safety), critical issues, priorities, and any help needed to be successful.

    a. Shift change: maintenance of outgoing shift to maintenance of oncoming shift
    b. Shift change: production of oncoming shift and maintenance of oncoming shift
    c. Maintenance personnel on shift
    d. Maintenance personnel with operations within shift (during a breakdown call)

2. Standard Process for maintenance staff work assignments
    a. Work priority is set via a process or a standard by which everyone abides.
    b. Work by crafts on shift must have a process to be leveled between employees.

c. A process exists to flex up and down reactive work resources with limited impact on productivity and planned work.
d. An overtime approval process exists.
e. A formal system of improvement exists, which includes: PM, PdM, problem-solving process and OEE teams (with operations and maintenance representation). The system is sustainable in good and bad economic times. The system produces results far greater than costs and therefore a good investment.

3. Reliability must be owned by operations. This is a foundational concept in reliability best practices. Operations makes the decisions on how equipment is run, when outages take place, where resources are assigned, and often how much money we spend on equipment. These facts make Operations the owner of the equipment and thus the owner of reliability. I know this concept can be controversial. It is argued among R&M professionals. Nevertheless, we all agree Operations plays a critical role. We will educate all employees in reliability in a later action item.

   a. Operations leadership will invest in planned downtime for maintenance to maintain equipment.
   b. Operations leadership will consider PM and PdM follow-up actions as well as problem-solving solutions when creating weekly production schedules. Fixing problems in a planned manner is understood by all to be far more efficient and less costly.
   c. TPM is to begin in Operations.
   d. A standard process exists for operators to support crafts during downtime events (planned and unplanned).

4. All employees feel empowered, hopeful, and know our CoatCo processes to improve.
   a. OEE teams exist

b. New roles focused on improvement are in place, aimed at improved worker skills

From this Target State, the remaining three hours were spent detailing actions that would bring the Target State to reality. Since we all understood the Target State, Current State, and real shop floor observations, these too flowed pretty quickly. We addressed each Target State condition as if they stood alone. Then we would look at all actions in totality to see if we had any duplications or synergies where we could combine two or more actions into one. The next chapter details our final cut of actions after our three hours of input.

Chapter 5
Actions to Achieve Target State

The actions below to achieve Target State have been expanded for your understanding. I chose to use bullets because I thought it flowed better.

1. Standards for Communication: Target date to implement: 30 days. We decided this was a simple but impactful change. It really just requires documentation, a communication plan, and an audit process.

   - <u>A maintenance communicator position is to be created</u>: There will be one maintenance communicator for each shift, for a total of 3. This person comes in 60 minutes prior to their assigned shift. They will meet for 30 minutes with the outgoing shift communicator to discuss current issues and necessary follow-up work. Next, the oncoming maintenance communicator will participate in the pre-shift production meeting detailed below. The outgoing maintenance communicator's shift is over after the 30-minute meeting with the oncoming maintenance communicator. The net result is 30 minutes of overtime per day for each maintenance coordinator.

   - <u>Pre-shift production meeting</u>: We currently have one shift leader on each of the three shifts. Minimum meeting attendance: two production leaders and a maintenance communicator.
     a. Meeting begins 30 minutes prior to shift end

b. Agenda:
     - Review EHS and customer issues from last 24 hours
     - Review critical production KPIs from previous 24 hours
     - Agree on priorities for next 24 hours
     - Ask if anyone needs help to achieve next 24 hours' goals. The production manager is first call on the help chain. The plant manager is second.
  c. Standard format for each shift. The same KPIs and same meeting agenda will be in place for all meetings. The production manager will own the format. Once we test the format for two weeks, we will build a whiteboard with rows and columns to further standardize the process.
  d. The production manager owns this process, including all changes.

- **Maintenance shift meeting:** Start of each shift (immediately following the production pre-shift meeting). The maintenance communicator shall lead the discussion. The meeting is attended by all shift maintenance personnel.
  a. Agenda:
     - Review EHS issues from last 24 hours
     - Review critical production KPIs from previous 24 hours
     - Review critical maintenance KPIs and asset performance from the last 24 hours
     - Agree on priorities for next 24 hours
     - Ask if anyone needs help to achieve next 24 hours' goals. The maintenance manager is first call on the help chain. The plant manager is second.

- Ask team if any short-term planning/kitting is needed to reduce downtime expected in next 48 hours
b. Standard format for each shift. The same KPIs and same meeting agenda will be in place for all meetings. The maintenance manager will own the format. Once we test the format for two weeks, we will build a whiteboard with rows and columns to further standardize the process.
c. The maintenance manager owns this process, including all changes.

- <u>Maintenance weekly meeting:</u> Occurs each Tuesday on all three shifts. The meeting is attended by all shift maintenance personnel. The new roles of reliability engineer, PdM technician, and maintenance planner are to attend all three meetings. (These roles will be discussed below.)

  a. The beginning of shift meeting shall be extended to a total of 30 minutes.
  b. Agenda
     - First 12 minutes: The same as daily maintenance shift meeting. This is led by the maintenance communicator.
     - Next 13 minutes: Review weekly and monthly KPIs, project status updates, trends (on focus areas like roller bearing failures, unwind/rewind kickouts). This is led by the reliability engineer (a new role detailed below).
     - Last 5 minutes: We will trial "5 Why Tuesday" – all will participate in the root cause of a current issue. Ensure simple yet meaningful problems are selected. Purpose is both to solve and demonstrate the tool. This is led by the new role of reliability engineer.
  c. The maintenance manager owns this process and is first call on the help chain.

- Emergency Work Request: A standard script is to be followed when operators request emergency work. The production manager owns this process.
    a. Operator calls for maintenance on the radio
    b. The operator communicates where on the line to meet maintenance. A site should be selected to best communicate the problem together live.
    c. The operator remains with maintenance personnel to assist with repairs and learn the root cause. This is a major change and will be detailed more below. This step may take 75 days to implement. See "pay for skills" section in Chapter 6.
    d. Maintenance is to stay with the operator to validate the quality of repair or mutually agree on a quality check.

2. Maintenance work assignments: Target date to implement: 60 days. Critical factor: time to interview, test, and select people. The maintenance manager is to serve as the coach for development, process improvement and approval, and questions/concerns.

- Emergency Crews – 6 total persons, 2 on each of 3 shifts
    Chalk circle showed 2 persons can handle the unplanned workload 95% of the time. Problems are to be documented by the maintenance manager for review. Overtime is the first lever for problems (this process is defined below). We are implementing 3 more actions which are expected to further lower the workload on shift:

    1. Create and execute PMs and PdM routes;

    2. Create a new off-line mechanic role on day shift to address the vast majority of off-line issues that do not impact line uptime; and

3. Create dedicated problem solving (a full-time reliability engineer). Performance will be reviewed daily in shift meetings, and we will make adjustments as necessary.

The primary focus of the emergency crews is to keep the coating lines running. They are expected to work as a team and, as required, move between lines to minimize downtime and balance work. Each crew member should still have a line of primary focus. This focus will not only aid in gaining intimate knowledge of the equipment, but also drive a partnership with operations.

Off-line equipment issues are to be deferred to the day shift off-line mechanic unless there is an imminent EHS issue, or the issue is likely to cascade to line downtime. Minor actions can be taken if time permits, but ensure you report events and actions to the off-line mechanic for documentation.

- Maintenance Planner – One person on day shift.

    We are going to shift to a system of PMs, PdM, and problem solving to reduce our reactive maintenance. A planner will be owner and historian for all work executed by maintenance – planned and unplanned. This role will function as the planner, scheduler, job kitter, job stager, and crib manager. To get things started, together with the reliability engineer and PdM technician, this role is to lead the execution of an asset criticality exercise – followed up closely with PMs and PdM routes on the most critical assets. We suggest you start by interviewing crews on problems, since we have no historical records. My challenge to you is to target PMs to begin two weeks after you are assigned to the role, and PdMs to begin within 2 weeks of Level 1 certification on a technology. It won't be perfect, but it's a start.

- Reliability Engineer (RE) – One person on day shift

    This role of the reliability engineer is to solve problems. Using data, input from others on the team, and critical thinking, this role is dedicated to improving. The RE will analyze data and trends and craft corrective actions both as an individual and as part of a team. This role will lead and facilitate OEE meetings mentioned later in this chapter. The RE is to get Level 1 certified as a PdM technician in both vibration and lubrication. In doing so, the RE will be a partner with the PdM technician. Initially, this role must partner with the maintenance planner role to get our proactive program started.

- PdM Technician – One person on day shift

    The PdM technician will focus their efforts on condition monitoring of our assets based on asset criticality ratings. The role will be required to get Level 1 certified in infrared, ultrasonic emissions, and lubrication. Target to be certified is six months. Lubrication certification is also a requirement for the reliability engineer. This is on purpose, for we expect lubrication to emerge from problem solving as a critical skill we need at this plant. The PdM technician and RE are to share an office for synergies. I have found PdM resources combined are more impactful than the sum of individuals. Early on in this assignment, the PdM technician is to partner with the RE and maintenance planner to conduct a criticality analysis and get the proactive maintenance program off the ground.

- Off-Line Equipment Mechanic – One person on day shift

    This role will become the process owner for all off-line equipment. As such, this role will partner with the planner to create PMs and with the PdM technician to create condition monitoring solutions to align with asset criticality and best practices.

- General Guidelines for these new day shift assignments
    a. Roles will be selected by interview and testing. (You may have to use seniority if you have a union, but still require competency testing.)

b. Roles will receive a performance and competence review by the maintenance manager every 6 months.
c. Vacation relief for the emergency crews will be split among these four positions. The target is to have each person take one day of an emergency crew person's one-week vacation. The fifth day will be on overtime. In this manner, each person will only miss one day of work in their normal assignment.

3. <u>Overtime approval</u>: Target date to implement: 2 days
   a. For the first 90 days, all overtime is to be approved by the maintenance manager. This is so I get a feel for the need and balance with business impacts.
   b. The process is to call me if need arises between 6 a.m. to 10pm. For 10 p.m. to 6 a.m., make the decision and email me your justification. I will coach as required.

4. <u>Execution of planned work</u>: Target date to implement: 75 days. Critical path will be getting emergency crews in place, new day shift roles filled, and to have all personnel trained.

   Critical focus areas:
   a. PMs, PdM follow-up, repairs, root cause actions, and improvements
      1. Every Wednesday, one of the two coating lines will be planned down for a four-hour outage.
      2. Outage will be timed with a coating change (reducing the impact to 3 hours and no additional scrap generated).
      3. Outage start time: We will target outages to be the first four hours of day shift or the last four hours of day shift. This enables midnights and afternoons emergency crews to support the outage on overtime. Start times to be owned by production.

4. Crewing for planned outages:
   4 (days shift roles) + 1 Emergency crew + 2 Overtime = 7 persons
   (1 Emergency crew to cover the running line and off-line equipment)
5. An outage planning meeting will be held each Thursday. The goal of this meeting is to discuss the next four weeks' outages. The outage tasks and schedule will be set in this meeting, one week out. Weeks two, three and four outages will be discussed albeit briefly. The latter weeks will get more granular as we get efficient at planning and executing outages. The planner will draft and review outage schedules to be approved by operations and maintenance managers. The maintenance manager owns this process as well as improvements.

b. Get a Computerized Maintenance Management System (CMMS): A CMMS is critical for asset history, planning, scheduling, and improvement. The maintenance planner shall partner with the maintenance manager to secure funding, for vender selection, and for implementation. Implementation of a historian is critical and must be completed early in the change effort to not duplicate work. For example: We don't want to create an Excel version of our PMs and then have to create a new one for our CMMS. Lastly, the goal is to get all maintenance work into the CMMS. This includes unplanned work executed by the emergency crews. Priority will be given to get PMs and PdM in place, followed closely by unplanned work.

5. <u>Reliability ownership by operations</u>: Target date to begin implementation: 30 days unless otherwise noted. This work can be completed independently of other actions.

a. The production manager is to drive both the greeting change and the requirement to assist with repairs to the operator job expectations.
b. Operator classification to be trained on the expectations of "assist" with repairs.
c. All three production leaders to be trained on TPM. Target date: 60 days from plan rollout.
d. TPM "clean to inspect" events will take place during each weekly outage. The production leader on day shift will own the events to be executed during the weekly outages.
e. A tagging system will be implemented within 30 days of production leaders' training. Tags are readily available through Amazon. Operators will place tags on equipment that is operational, but is operating poorly, leaking, or showing early signs of failure. Tag status will be reviewed during the extended Tuesday maintenance staff meeting.
f. The production leaders on afternoons and midnight shifts are to implement 5S on the coating lines. Together they can decide how to divide up the work between the two shifts. The only rule is to make every effort to keep the standards identical on both coating lines.
g. Operators are to attend a biweekly OEE meeting (see below for more information on this).
h. This entire process is owned by the production manager.

6. <u>Problem solving</u> – A system for improvement: Target date for implementation: 90 days.
   a. Start OEE team for downtime on each line.
   b. Meet biweekly for 1 hour after each shift (3 meetings).
   c. Include operations and maintenance personnel.
   d. Agenda, prework, and overall ownership of this meeting belongs to the reliability engineer.
   e. Agenda to include:
      - A review of reliability events for the last month and rolling 12 months.
      - A Pareto chart of events

- Updates on selected focus areas. Example: Unwind kickouts.
- Input/discussion

f. The maintenance manager will own this process.

7. <u>Pay-for-skills</u>: Target date for implementation: 75 days. Critical path is to identify successful candidates for day shift assignments.

   a. PdM Technician: $2 per hour pay premium – 1 person (Estimated cost: $4K/yr)
   b. Maintenance Planner: $2 per hour pay premium – 1 person (Estimated cost: $4K/yr)
   c. Reliability Engineer: $2 per hour pay premium – 1 person (Estimated cost: $4K/yr)
   d. Line operators (3 per line per shift = 18 total) to assist maintenance with repairs – $1 per hour pay premium (Estimated cost: $36K/yr)

Pay-for-skills may be a difficult pill to swallow and is not required for this plan to be implemented. However, it would create significant momentum and organizational enthusiasm. By implementing pay-for-skills, management is recognizing the contribution of the team members who have stepped up their skills for the mutual benefit of all. Further, it gives hope to others in the organization by providing roles to which they can aspire. Another option would be to set up a pay-for-performance plan which awards employees for hitting meaningful milestones. I have had limited success with these programs over the long term. They tend to evolve into profit-sharing plans since individuals feel powerless to make a difference by themselves. In addition, I've found that over half the time market factors have a bigger influence on results than individuals. This can lead to distrust in management, as they could potentially manipulate numbers to impact payout. Pay-for-skills also shows management's trust in their workforce and allows individuals to excel regardless of the performance of others or the market. This action is to be owned by the HR manager.

8. <u>Audits</u>: Target date for implementation: 30 days
One lesson I've learned in 33 years: if a process or action is critical, you need to audit it. Every action listed here needs to have an audit process developed to ensure compliance and to prevent drift. "Drift" is the slow evolution away from the standard over a long period of time. Audits can be a simple checklist for critical content where a score is given and reported. This audit is then audited as part of an assessment of the plant. For example, Jeff, the production manager, is to audit his pre-shift production meetings. This is a simple check sheet on the following: attendance in meeting, meeting started on time, meeting ended on time, agenda was followed, all attendees participated, and meeting judged as effective. This meeting is to be audited weekly. The results of the weekly audits are then reported to the quality manager who audits compliance for the whole plant on dozens of standards. The quality manager reports the compliance of all plant standards to the plant manager monthly. Once a quarter, the quality manager sits down with Jeff to review compliance one-on-one.

9. <u>Communication, selling, training, and reinforcement Plan</u>: Begins today and continues indefinitely.
I put communication, selling, training, and reinforcement last, not because they are lowest in priority, but rather for retention – to make sure they are the last actions you read about so they'll stick in your head. You will fail if you don't dedicate time and resources to this step.

Communication: A step-by-step plan with timelines needs to be created for telling employees about the above changes. In that communication, ensure you have time for questions and feedback. Next, how are you going to continue to reinforce the plan after initial communication meetings? weekly email blasts to the organization? posters? bulletins? monthly reports? Ensure the individuals who participated in the chalk circle observation process get one-on-one communication and a "thank you." What is expected of leaders in selling this change? Should they give the program an endorsement in every meeting right after EHS updates? Every member of the lead team has a role in this step. My recommendation is to have your HR manager lead this action. At CoatCo, the lead team did in fact elect Jordan, our HR manager, to lead this phase of the project.

Selling: If you are in reliability and maintenance – and especially if you are driving any change – congratulations, you have been appointed an honorary sales manager. Human nature makes it easy to talk about the negatives of any change as well as gossip about what could go wrong. Selling takes personal confidence and leadership. Selling overlaps quite a bit with the communication plan; thus, the HR manager will lead this effort as well. As stated before, I like to sell controversial changes as 90-day experiments. This is a contract you make to learn, adapt, and reconsider the changes given new information.

Training: A 2-hour course in reliability basics will be created and given to all operations personnel. Maintenance personnel will receive 4 hours of reliability training. I can design and present this training at no cost.

Reinforcement: How are we going to reinforce our changes? Some ideas to consider after you've achieved milestones in implementation or results:

- Celebration meals. For example: We achieve under 10 hours of unplanned downtime in one week. Recall our average is 21, and we have never been less than 15 hours.
- Buy shirts commemorating our new approach to reliability. The print may say something like: "CoatCo – Together Driving for Excellence." Make the shirt something they can wear to work.
- Buy $75 gift cards for all employees to take their spouse or significant other out to dinner.

- A personal handwritten note from the plant manager to each person who took on a new role or made a significant impact. Buy cards with a nice saying on the front and have the plant manager write about specific actions completed by the individual and how they connect to the business on the inside. Make the note "refrigerator worthy" for the individual. This is huge and almost cost-free. I set a goal for myself to write 4 per month.

The difficult side of reinforcement is how leadership addresses employees who refuse to adopt the new standards of work. This is emotionally challenging, but the whole organization knows Billy is not following the new expectations. My advice is to always coach one-on-one first with the direct supervisor. Ensure there is not a skill or other legitimate gap. If the problem ends up being that Billy just doesn't want to play, Billy needs to advance to the exit lounge. Not addressing employees that refuse to accept new practices is a major problem at most companies. Many just develop workarounds, but this ignores the negative impact the underperformer has on others. I address this later in Chapter 13, "Mistakes."

There is a popular change model called ADKAR by Jeff Hiatt. The model name, as you'd expect, is an acronym: Awareness, Desire, Knowledge, Ability, and Reinforcement. Each of these components must be addressed for successful change. Notice how going through the A3 and our actions hit all aspects of the model except reinforcement. I believe the A3 process focuses more on what needs to improve, and ADKAR focuses on how to help the people accept the required changes. Nevertheless, reinforcement is a critical component of change and should not be skipped. If you want to learn more about the model, there is a lot on the internet.

This is the end of your actions to achieve Target State.

Points to Ponder

1. Notice how these actions compare to our proposed actions after the business case in chapter 2; these two groups of actions could not be more different, right? Observation completely changed our understanding of the business and thus changed our leverage points.

2. Notice how fast these actions can be implemented; the longest timeline is 90 days, and several are just a few days. Culture changes and results should be noticeable within a month.

3. On a scale of 1 to 10 (with 10 indicating 100% chance of success), how do you judge the probability of success in the CoatCo culture for these actions taken from Current State versus those after the Business Case alone? I'd rate the actions from Current State about a 9 – high probability of success. Actions from the Business Case alone, I'd rate a 1 – very low probability of success.

4. Which action list do you think cost more: after Current State or after Business Case alone? No question – the actions identified after Current State will cost a fraction of what those from the Business Case will cost. In the next chapter, I will break down the cost and business impact.

5. In summary, you just learned how gut reactions or drawing conclusions from your past experience alone most often leads to poor actions, higher costs, and a low probability of success. Observation changes everything.

Chapter 6
Business Impact

From the Business Case, recall our "mile marker" for success is the 12-month run rate. That is, at the end of 12 months, what are our projections based on demonstrated performance of future months? When it comes to financial projections, we are not looking for perfect. In most financial forecasting, controllers are looking for a reasonable logic path where there is an equal chance results could be higher or could be lower. I tend to put my thumb on the scale to ensure I under promise and over deliver; the numbers below reflect this bias. I drafted these impacts myself for a later review with the lead team.

Anticipated business impact – Run rate after 12 months

(Recall the plant operates 50 weeks per year)

1. Production – I am projecting maintenance downtime to drop from 35 hours per week to 20 (including the weekly 4-hour planned outage). Using the same one hour of downtime valued at $500, the positive impact is $7,500 per week, or $375,000 per year.
2. Scrap generation – Recall every downtime event results in 2,000 feet of coated scrap at $0.25 per foot (lost revenue and remelt cost of scrap to recover the base metal – the applied coating is not recovered). We currently average 21 unplanned maintenance events per week; we expect this to be reduced to 11. This is worth $5,000 per week, or $250,000 per year.
3. Quality – Current state customer returns for equipment issues related to reliability is $20,000 per month. We expect to halve this to $10,000, or $120,000 per year.
4. Overtime Controls
   - Observed 2 hours of overtime per day per craft.

- 10 craft x 2 hours/day x 5 days in a week x 50 weeks = 5000 hours/year
- At $30/hour base pay, overtime is paid at 1.5 times base, or $45/hour. Therefore, current state overtime is **$225,000/year (reduction)**.
- Planned outages required 2 crafts to work 4 hours of overtime per week, or 8 hours.
- 8 hours x 50 weeks x the same $45/hour = **$18,000/year (increase)**.
- Maintenance communicator 30 minutes/shift overtime = $17,000/year (increase).
- Full impact of controls: $190,000/year (reduction). I want to employ a 50% factor here and round up to $100,000/year impact. At least in the first year, I want to be cautious with overtime reductions.

5. The sum total of these savings:

    $375,000 + $250,000 + $120,000 + $100,000 = $845,000 per year

Cost estimates of these changes

- PdM certifications for the PdM technician and reliability engineer:
  $1000 per class tuition + $1500 hotel, meal, and travel expenses to attend training + 5 certifications (UE, IR, Lube (2 persons), Vibe) = $12,500.
  This is a one-time charge incurred in year one only.
- Lease of PdM equipment: $10,000 per year.
- CMMS: $50,000 initial setup charges (one-time charge in year one) and $12,000 licensing fee per year. This allows for 20 system users (Estimate: $50/user/month)
- Pay-for-skills: $48,000 per year
- OEE meetings on OT: $8,000 per year
- **Year 1 Total investment: $140,500**

- **Yearly investment beginning year 2: $78,000**

Net Business Impact – Run rate at 12 months:

- ✓ $767,000 per year savings ($845K - $78K)

- ✓ While this is the run rate at 12 months, savings should begin in 30 days and escalate through year-end. I estimate the year one savings to be near $300,000, which is more than double our cost. Consequently, this plan solves the plant manager's challenge to have quarter-over-quarter improvements.

You may have noticed I did not claim any savings for parts and materials. My experience tells me these costs will go down since we are doing PMs and PdM routes. However, it is important to realize we will find equipment that is in need of repair but has not reached final failure. Consequently, some costs will be frontloaded. In my experience, I've found the savings from our proactive work will offset many of the repairs we find. We will have to monitor this to see if this situation is repeated at CoatCo. Nevertheless, there is no way of calculating a savings or a cost for "we don't know what we don't know." Worst-case spending will eat into our first-year savings. Regardless, whatever happens with parts does not change our actions toward achieving the Target State.

Other Benefits

1. Engagement – We have a simple plan that provides a formula to win and provides hope to the organization.
2. There is a direct correlation between unplanned work and injuries/environmental issues at plants. Research on this topic can be easily found with an internet search.
3. Attrition – Crews will learn new skills, gain increased base pay, and become an integral part of the success of the team, resulting in personal pride and accomplishment.

4. Opportunity cost associated with poor reliability – There is an incalculable impact on the organization from being consumed in emergency work. In a reactive culture, resources are mentally and physically exhausted just trying to survive. Even in unexpected times of good reliability, the staff is reluctant to start any improvement projects because they know this is just the lull before the next reliability storm. Examples of opportunity cost: What is it worth to have resources focused on line speed, quality, inventory reduction, and new products?

Risks and Concerns

Every change is hard regardless of how simple the actions may seem. Every change also comes with risks and concerns. I'll try to read your mind on what these may be for you at your site:

- "My plant does not have the right skill set for a planner, PdM tech, or reliability engineer inside the current maintenance staff."

    Advice: Do the best with what you have at the beginning. The maintenance manager will need to spend a lot of time coaching these folks. To give them confidence, start with the areas in which they excel. Quite often, pushback and challenges are really outcomes of fear. Training reduces fear; therefore, consider outside training. The candidate may not be perfect, but ask yourself – would the organization be better off without the role, lost again in the inefficient, reactive maintenance world, or are you better off with them in this proactive, planned role? The answer should be obvious; planned work wins every time. If anyone leaves the maintenance organization, hire the correct skill set. If no one volunteers for the new role out of fear, nudge someone to take the assignment for 90 days. Ensure they know that if they fail, you will place them back in the emergency crew.

- "Joe, Joe, Joe – your savings estimates are way off."

Advice: I stand by my statement that I have under promised and will significantly over deliver. Nevertheless, for a moment, imagine I'm off by 50%.

So the math is: *50% x $767,000 = $383,500.*

This exceeds what the lead team said in the business case would be "excellent" in 12 months. So, being rated "excellent" is our worst-case scenario – I'll take it.

Secondly, in project analysis you must always compare proposed changes with the "do nothing" option. It is foolish to assume that doing nothing will lead to the same results you are getting now. Looking ahead in time, there are many negative pressures to consider, including inflationary costs for all materials and services purchased by the plant, employees' desire for annual pay increases, health care costs skyrocketing, and aging assets getting older and older. With this aging, you have increased sustaining cost and increased risk of catastrophic failure, including major business interruption. "Doing nothing" is the high-risk action in my experience. Have your controller help you with this.

Lastly, recall the cost of the previous maintenance manager's plan. He was calling for a year one cost of $1,400,000 with no promise of results until years three to five. Further, the results promised were equal to costs – a pretty bad investment. Our A3 action plan is starting to look pretty good now, right?

- "The plant lead team does not agree with the proposed actions."

Advice: This can occur when one or more lead team members are solely relying on data, KPIs, and previous experience to drive their decisions. While this does not sound bad, these folks are outright disagreeing with the Current State of this A3. Consequently, you need to sell them on the process and convince them that Current State must supplement data, KPIs, and experience. I have two suggestions for you here:

1) The quick and easy solution is to revert back to the raw observation data, review common themes, then show how the actions address the common themes; and

2) Take the doubters on a chalk circle observation of current state. This may take you two days, but you must get them to be passionate leaders of the changes coming. If either of these don't work, you may have an employee problem, so I would have a private conversation with the plant manager. If you deem alignment of the lead team to be a major concern at the outset of the A3 process, alter the observation process to include more or all of the lead team. Observation always trumps opinion. In this manner, they will get overwhelmed with observed waste and emerge as supporters. I have never seen this fail – never.

Chapter 7
Measures of Success

I drafted the measures of success for later input and approval from the lead team. Each measure needs to have a lead team assigned to champion it. The top five measures were the ones selected by leadership as our key performance indicators (KPIs) for this A3. The remaining 13 are all drivers to these five. At CoatCo, we assigned owners to all 18 measures but were only expected to report on the KPIs. Warren Buffett has a great philosophy on this called his "5/25 rule." It's easy to look up on the internet, but in a nutshell, it states that dramatic results come from focusing on the critical few. I have more on the 5/25 rule in the quick reference section at the end of this book. The second set of boxed measures are all audits; these were detailed earlier under the audit action item.

Measures of Success

1. Sales per month versus baseline
2. Hours of unplanned downtime in maintenance per month
3. Dollar value of downtime per month (including scrap value)
4. OEE of Line 1 and Line 2
5. Dollar value of quality returns per month
6. Number of 4-hour planned outages per month versus plan
7. Total PM hours per week
8. Planned work percent for the month
9. PdM cost avoidance dollars per month
10. Problem-solving cost avoidance dollars per month
11. Total number of unplanned overtime events required to restore flow, per month

12. Audit of 5S at lines
13. Audit of craft – operations connection (person to person)
14. Audit of shift meeting – maintenance
15. Audit of shift meeting – operations and maintenance
16. Audit of OEE meetings
17. Safety performance
18. Attrition

The last step is to review business impacts and measures with the entire lead team. This meeting will take about 90 minutes. I like to send my draft to participants as soon as possible before the meeting. This makes the meeting much more efficient. Other than clarifying questions, my meeting with the CoatCo lead team ended with full buy-in for the new plan.

Congratulations! We have now completed the A3.

Chapter 8
Reflection – What have I done?

The plan we just created is targeting a change in the culture. By creating new experiences, we move people from powerless to powerful. You will not only see results from these actions, but you have also established an open learning culture that will enable future actions to take root. The actions yield meaningful improvement to the business in just weeks, negligible upfront investment, and no bow wave of spending before positive results are realized. Further, this process increases organizational hope and unlocks employees' hidden talent. Employees will see a more enticing future due to opportunities for skill development and increases in pay. All these actions will drive lower attrition and improved safety performance. Not bad for 30 days.

What is the key to such a turnaround? Answer: Thorough understanding of current state revealed through intense (chalk circle) observation. Without observation, most organizations implement actions created out of the business case alone. Actions, which we just showed through example, were horribly wrong, costly, and wasteful. Plain and simple: observation enables the leader to make better decisions.

A traditional reliability deployment costs significantly more, delays results, and has a 95% chance of failing due to the loss of sponsorship or actions getting eaten alive by your culture. Plus, the wastes we unveiled during observation would remain. Why would anyone sign up for a traditional deployment? Answer: It has worked in the past. Which is a good answer if all factors of your business climate have remained the same – which they have not. The expectations put on plant managers and all leaders are vastly different today. Perhaps the largest of these is the expectation to produce positive results quarter over quarter. This requirement dooms the traditional deployment unless you have a very unusual leader.

How do you expect Jane, the plant manager, to respond to this A3 proposal? Obviously, given the way we laid out the A3, wallowed in the waste of current state for an eternity, and had our actions attack the wastes with results far beyond our expectations— all for no upfront cost— she will be both delighted and inspired. Delighted in the results, and inspired by the new process she has learned to run her business: the A3.

Why do the majority of leaders and change agents not embrace observation as a tool to understand current state? I firmly believe the pace of our lives and jobs has resulted in us "cutting corners" to create the illusion of increased efficiency. We cut corners with our family by not having dinner together around the supper table. We cut corners with our friends by sending a text as opposed to having a live conversation with them. Leaders cut corners at work by sending an email versus having a meeting to debate. Email has taken over an increasingly large part of our work life. When my career started in 1987, my company did not have email. We had paper mail, but very little of it. The time I spent reading letters was insignificant, perhaps 30 minutes a week. Today, how much time do we spend reading and responding to emails – two or three hours a day? That's a massive commitment. What has been edged out of your workday? You guessed it – time out in the process, seeing reality. Technology gives the illusion of being connected. We see automatically generated KPIs, attend meetings, and hear opinions; from these, we conclude we know what is going on. As this case study demonstrated, we have been fooled. I have hundreds and perhaps thousands of examples where observation revealed a different current state than I and others expected. Don't get me wrong – technology has a massive role in manufacturing, but when technology results in the total exclusion of shop floor observation, we drift from reality. I have never been to a plant with poor reliability that was due to an automated technology gap. Technology can be a great accelerator, but it's an accelerator from a solid "go and see" reliability culture. I believe most look at the shiny new tools of technology and believe they can skip all that culture stuff. They fail 100% of the time; if you choose this path (and many do), know your odds of success are pretty low.

An eye for waste and a relentless focus on waste in your culture and systems is the secret sauce, folks. It is critical to know that all reliability tools are designed to eliminate waste. Therefore, it makes no sense to apply tools without knowing the unique waste at your plant. Following the A3 process guarantees you stay on the waste path.

Many of you may look at this CoatCo story and resultant A3 and think that the actions and process worked very well for this plant with no use of best practices – but how would it work for a plant that has been on a reliability journey for 20 years and is using all best practices? My answer is, just as well. The actions will be different, but I'm 100% confident that the observations in the current state evaluation will yield significant waste. Having all the best practices in play says nothing about:

1. Wrench time
2. Precision maintenance
3. Communication within and between groups
4. Connection waste between groups
5. Effectiveness of problem solving
6. Effectiveness of sponsorship
7. Management decisions counter to a reliability culture

I have yet to see the perfect plant with zero waste. There will always be waste to reduce for those that are trained and choose to see it.

Chapter 9
Next Steps

The A3 actions we took will not be perfect; you will need to adjust. Make sure you adjust with new information that <u>INCLUDES</u> new observation. The key for you is to start; don't wait for a perfect plan. In 6-12 months, you will need to create a new A3. When, exactly? Time the new A3 event for when you believe the previous A3 Target State has been realized and stabilized into the new normal. The A3 process is repeated indefinitely; it is your process for all change.

Now that we have a foundation, a new culture, and established creditability, what can take us to another level? Remember, we have created enthusiastic sponsors. Perhaps…

- Introduce enhanced skill training to go deeper into planning, PdM, or problem solving
- Hire a skill set we don't currently have, like a highly skilled reliability engineer that would bring enhanced analysis and tools to our plant
- Pursue new technology, such as continuous condition monitoring and artificial intelligence on critical assets
- Drive for world-class lubrication. Fund a lube room, label all assets with grease and oil types, install oil sampling ports and sight glasses, add desiccant breathers, and add automated greasing.
- Transfer learnings to sister plants within your same company.
- Other ideas?

We are saving $767,000 per year and have a system to improve even more. We have earned CREDITABILITY, folks. Let's cash it in on investments that will take us to higher levels of performance. Ensure these actions that require investment will create a new A3 Target State. Do not install solutions in search of problems, but rather craft solutions to known problems.

Chapter 10
Advice

If you have enjoyed this book, take a look at my YouTube channel titled "Reliability Man." I have over 80 videos on driving results in your business by excelling in reliability and maintenance through lean principles. I make a new video weekly. I respond to all comments and questions. Also, feel free to reach out to me via my email: leandrivenreliability@gmail.com.

I'm a very seasoned leader in reliability and maintenance. I realize I made the waste and interventions at CoatCo seem obvious and apparent, but that is what an experienced guide can do for your company. If you are beginning a reliability journey, hiring an experienced guide will shave years off your deployment and save you money.

The process detailed here is scalable to any size plant. At larger plants, I conduct observations by forming and training teams of observers in order to get a sample size large enough to represent the plant. Typically, a team consist of two persons: a lead team member and a coach (both from host plant). The coach I work with one-on-one before observation days. The teams report back to the larger group with their observations, common themes, and hypothesis. This is the same basic process I use at smaller plants like CoatCo. Using the team approach has the added benefit of helping lead team members see the value of observation by taking part in the process. It is my number one tool for getting lead teams excited about reliability. I've used this team observation process about 90% of the time. Plant size is a major factor in this decision. The biggest challenge for the team observation process is finding time for leaders to fully participate in the A3. It is essentially a one-week commitment that will change how they lead; still, it is difficult to find the time. In all honesty, this is disappointing, but they usually get inspired when I show the results of my observations. Inspired – but not to the same level they would have reached if they observed themselves. They also miss out on learning the skill of observing waste. This skill should cascade down to the whole organization, so full team involvement is highly preferred.

Targeted advice for plant managers

Hire an experienced guide. How is that for simple? Do not accept a solution that has high up-front cost and a promise of future savings; that's the 1980s approach to reliability. Refuse the boiler plate solution; a one-size-fits-all deployment will be 90% wasted effort since all the actions do not solve the issues unique to your plant. Don't assume reliability is complex. Great leaders are great simplifiers. You can do this. Lastly, strive to create a "go and see culture." KPIs are just the start. Observation of reality is empowering and will result in better decisions and better, executable actions. Warning: Observation is a skill that needs to be developed. While it appears simple, a trained guide can see far more waste than a novice, primarily due to bias. The novice frequently discounts waste as an unusual event or in fear of embarrassment; they don't want to highlight the poor practices of groups within their area of accountability.

Targeted advice for production managers

First and foremost, a production manager needs to comprehend that they are the owners of asset reliability. Production leaders make the decisions on how we run assets, how much time maintenance spends on the equipment, when maintenance gets the equipment, when and if new equipment is purchased, how much maintenance spends, and where resources are focused – production leaders are clearly the "deciders" at a plant. Maintenance leaders are "advisers." They give guidance on actions that will preserve or enhance reliability. They guarantee that work is performed efficiently and problem solving occurs – but this is an advisory role. Imagine the synergy that is unlocked when you have production, maintenance, and engineering all aligned on reliability. This is where excellence emerges. Unfortunately, most organizations settle for conflict by having production focused on output and maintenance focused on reliability. Expect better.

Secondly, data-based decision-making is key to your success, but data does not end with KPIs, charts, and graphs. Chalk circle observation is a must to be successful. Don't bypass it. Observation is how you know current state and avert bias. If you don't think data is biased, do some research into the assumptions and calculations made to collect the data. Here is a simple example of hidden waste in a KPI:

Setup time to change products averaged 30 minutes last week. The last 12 months rolling average is 32 minutes. Good week, right? Well, what if I told you the individual set-up times, in minutes, were 18, 41, 23, 31, 42, 35, and 20. There is huge opportunity in understanding how we achieved the 18, 23, and 20 minute turnarounds. There is equal opportunity in trying to understand the 41 and 42 minute turnarounds. Actually, I bet there is something to learn from each and every turnaround. That "something" is waste, and it's hidden in the KPI – but reveals itself in observation.

Targeted advice for maintenance managers

Be a leader. A leader risks failure. The vast majority of maintenance managers are overwhelmingly conservative and seldom take chances. Perhaps some of this is the nature of the job. You never get praised when everything goes right and get reamed out if anything fails. I've been there. Nevertheless, reliability tools are rooted in science and proven time and time again at plants. You need to take the leap. However, it is not a leap of faith; it is a leap of science. Yes, you will have some failures, but you're failing right now. Be bold. Lead.

Secondly, even Michael Jordan, argued by most as the number-one professional athlete of the last 100 years, had a coach. No matter your experience, a guide that has already created a reliability culture can help you get better results faster.

Targeted advice for a front-line supervisor

Help top leaders create a "go and see" culture. After an equipment failure, go inside the offices, grab a lead team member, and show them the failure up close and personal. Discuss the whys and solutions. It is all too easy to discount failures as a normal part of business in the conference room or in an email. It gets more real and more actionable in the field. This should be repeated for successes from root cause work or new installations. Give the leader positive reinforcement for the decisions they made.

Targeted advice for a maintenance planner

You own wrench time. Perhaps you've never heard this before. The top limit of wrench time is set by your critical planning function. Consequently, as part of a job plan, you must anticipate wrench time detractors. This can take many forms; here are four examples:

- Validate the pump you purchased has the right base plate and shaft diameter.
- Meet with the crew executing critical work during an outage a week ahead of time to ensure they agree on how to lock, tag, and verify equipment is at a zero-energy state.
- Set up a standard meeting with the crew supervisor prior to an outage to review your plan for special tools and support equipment.

- Insist on the creation of a job kitter/stager position(s) at your plant. Their role is to assemble everything necessary to perform the job in a kit and to stage the kit and all equipment required at the job site. There is no bigger and quicker driver of wrench time than having such a role.

Since you own wrench time, do you ever audit it? Does anybody? If the answer is no, you need to meet with your manager to see how this data can be attained. I strongly suggest using the chalk circle observation process. To combat fear of embarrassment, if at all possible, you should observe jobs planned by other planners. This will enhance your learning, since you are not emotionally attached to the jobs.

Targeted advice for a reliability engineer

You are in sales. The plant manager should not be expected to know the failure that did not happen. If you have extended the mean time between failure on a critical pump from 12 months to 12 years, you need to advertise it. This is not tooting your own horn; it is you earning creditability for the next change you suggest. I suggest a weekly email blast with a success of the week. I also like monthly newsletters detailing successes, failures, what's next, and help needed. Insist on face time monthly with the lead team to dive deeper into the content of the monthly report. Lastly, look for opportunities to take lead team members into the field to see failures and successes. This "go and see" culture will more firmly engrain reliability lessons into their thinking process. Reliability programs die from a lack of sponsorship. The best way to prevent this is by selling successes and education. The next business downcycle is coming; don't let reliability be on the chopping block due to ignorance. Reliability needs to be viewed as a waste elimination process, and during downcycles work should be accelerated. All this depends on you selling.

Targeted advice for a PdM technician

You need to help take the mystery out of the predictive maintenance tools. Leaders need to know this is simple science, not opinion or voodoo. What I suggest is that you invite – be persistent – each lead team member to join you on a route. Not a full route – ask for an hour. Have them use the IR gauge, put on the UE headset, or take the lube sample. When you get an anomaly on a route, come into the office and persuade a leader to join you in the field to witness the failure detection. Lastly, once a quarter, encourage your supervisor to set up a meeting with all the PdM techs and the lead team. Have the PdM techs present successes, failures, what is next, and what you need. This will become the highlight of the lead team's quarter; it was for me as the plant manager.

Chapter 11
Selecting a Consultant

Several times in this book, I recommended hiring a guide. While not a requirement, it will shave years off your deployment and front-load your results. I believe every consultant has something positive to contribute to your journey. However, most follow the same time proven model. If you don't know already, my process is a definite outlier. My hope is that you view it as an alternative; not better or worse, just different and perhaps the right solution for your plant.

The concepts I present here are not that complicated. Best practices like planning and scheduling can be easily found in books or a simple internet search. However, it's never the plan that destroys change efforts – it's the obstacles. Most of these obstacles are culture-related; consequently, a standard one-size-fits-all solution does not exist. Managing obstacles is where a consultant can really earn their fee. So, look for a guide that has experience actually implementing a reliability culture. Avoid consultants that have studied others or have learned concepts in a classroom and, hocus pocus, an expert is born! My strong recommendation is to look for someone who has both been the accountable person for the results (for example: a maintenance manager, reliability leader, plant manager, or vice-president) and has experience in consulting. Look for a multi-plant resume so they have a good bandwidth of experiences. Ensure that the consultant that shows up at your plant has this resume. Far too common is the "bait and switch." The project leader has the skills you're looking for (the bait), but a direct report of the project leader is the individual that shows up on your plant site (the switch). Lastly, look for a consultant that has the heart of a teacher. You don't need a hero to come in and fix your plant and then leave – or worse, fix the plant and stay (for enormous fees). Look for a guide the helps you craft and implement systems to both sustain and improve the plant. The plan must be owned and implemented by you. You must see a path to wean yourself off of a guide.

Lean Driven Reliability (LDR) – Your Experienced Guide

I know I'm biased, but LDR is the exact guide I describe above. I have 33 years' experience in manufacturing, with 28 in plant leadership roles including engineer, department manager, and plant manager. The remaining 5 years were in consulting – 4 years as a global internal consultant to Alcoa and one year with LDR. I have consulted at 32 global locations with dramatic results. R&M savings in year one was a minimum of 10%. These savings were systemic, rooted in best practices, and laid the foundation culture for future improvements. Production improvements always accompanied these results, but varied in nature. Keep in mind, the production and quality impact at most plants can easily be two times the R&M savings. The plant I retired from realized a 29% improvement in capacity on top of their 10% R&M savings – all in year one.

The secret to my process success is the application of lean concepts to reliability best practices –most notably, the use of observation and the A3 process to clearly understand current state before selling you tools. Imagine going to the doctor and getting the same surgery as the person before without an examination. Crazy, right? Well, be careful. This is what some consultants are selling. Knowing current state allows you to apply tools to your waste, as highlighted in the story of CoatCo. Results come fast, enthusiasm is achieved, and sponsorship is energized and complete.

I first heard the term "Heart of a Teacher" from Dave Ramsey, the financial advisor and radio personality. Some people are born with the skills and desire to help others succeed. This is what drove me to leadership roles my whole professional career. I spent at least 30% of my calendar time growing leaders. – not being a puppet master telling everyone what to do, but developing their leadership skills.

While at Alcoa, I started Alcoa University, which had over 70 leadership coaching classes for up-and-coming leaders. All classes were taught by seasoned managers. This design had three pillars for success:

1. It taught needed skills to new leaders
2. It facilitated existing managers to get better at the skill they were required to teach; and
3. New leaders were quickly surrounded by a network of leaders committed to their success.

Typical classes included "How to lead fatality prevention plans at a plant," "Reliability Excellence," "How to drive improved performance by a team," and "How to give a performance review." Classes were as short as 2 hours and as long as 5 days. Students signed up for classes under the guidance of their direct manager. Another feature of the university was the focus on proficiency. While a few classes were just "pass/fail," most classes provided a rating for students: 1 (not trained), 2 (trained), 3 (proficient), and 4 (subject matter expert). Proficiency was earned by demonstrating the principles taught in class to the professor out on the shop floor at least 5 times. The status of subject matter expert requires proficiency – and the student must begin teaching classes on the topic. The student and their manager agreed on the status to be achieved as part of their individual development plan. As the plant manager, I was the dean of the school and professor of 5 classes. This is a very low-cost training solution and, I contend, much more impactful than outside training due to the 3 pillars. Contact me if you would like more information on this topic.

Since retiring from Alcoa, I started the YouTube channel "Reliability Man." Each video explores a topic and provides a simple action the viewer can employ in the next week. My hope is to coach the masses on my process of creating a reliability culture using lean. Most viewers I will never meet or correspond with. The channel keeps me quite busy both creating videos and replying to comments and emails, but I just love the subject matter.

Contact LDR for...

- Reliability assessments with a focus on waste and best practices
- Reliability solutions and personal coaching – on your site and remote
- Your experienced partner and former plant manager qualified to sell reliability to upper management
- Industry speaker (recently presented at SMRP and MARCON)
- Author of *Zero to Hero: How to jumpstart your reliability journey given today's business challenges*

Contact Joe at LDR: leandrivenreliability@gmail.com

Role of large, traditional consulting firms with reliability deployments

If you compare my approach to other reliability consulting firms, you will see vast differences. My hope is to give you an alternative to consider for your journey. The traditional reliability deployment begins with a comprehensive assessment of tools and processes in place at your site. This takes about a week (depending on plant size). The consultant is looking for the tools and practices, not the impact nor the efficiency of their use.

Next, the consultant will begin training— general awareness training for the whole population, then moving to specialized training on such things as planning and scheduling, effective PMs and PdM, problem solving, work control, stores management, unplanned work management, governance, and measures of success.

Then, they will help you get best practice processes in place for your plant. They will not focus on how efficient and impactful the new processes are on your factory floor, but rather on whether you know and employ the practices. Note: This is the exact opposite of the lean approach I use and demonstrated in the CoatCo story.

Lastly, they will also help you with challenges and any questions you have along the journey. This approach has worked for decades; it is how I learned about reliability. Unfortunately, the realities of this deployment plan make it difficult for top leaders to embrace in today's business climate. Most notably:

1. Huge upfront cost in assessment, training, and coaching. For a medium-sized company, this easily approaches $1MM.

2. Nearly all consultants will explain a bow wave of spending that must take place on your assets to get them in maintainable shape. This is getting a majority of items repaired to OEM (Original Equipment Manufacturer) standards. I have experience with this being another $2-3MM per year for 3-5 years.

3. Results are back-end loaded. Tangible success comes from all elements of the system working together. This can easily take 3-5 years after beginning your consulting partnership. At this point, you already invested $5MM.

4. The plant may need to hire new positions in planning, supervision, and reliability engineering to hit target ratios. Full up costs are near $150K for each position added.

These four facts collide with the plant manager's expectations of quarter-over-quarter positive results. A large investment with the hope of return in 3-5 years does not compete well at all with other investment ideas brought to top leadership. Everyone knows that reliability is as much a culture change as it is a use of tools and best practices. What are the realistic odds your plant can complete an uninterrupted reliability journey kickoff for 5 years, given business cycles and other management agendas? Not good, right? Consequently, regardless of how the plant manager feels about the benefits of reliability and the resume of the consultant, they cannot endorse this approach.

My fundamental divergence with nearly all consultants is how to start a reliability journey. Rather than a standard "fix everything" approach, I believe plants would be best served by consultants who first focus intensely on understanding the unique waste at your plant, and then apply best practice tools with precision to your waste and inefficiencies. Consequently, results will be rapid and sustainable and fuel new investment for better and better results. All with little to no upfront investment by the plant. Good news -- you now have an alternative.

I fully support the work of large consulting firms coming into plants after two or three iterations of A3s like we completed in this book for CoatCo. These firms are very competent at creating systems and processes that are consistent across large plant sites (think 1,000 or more employees) and multiple sites. By establishing common systems and practices, improvements at one site can be rapidly transferred to others. Software tools needed as part of the deployment can be standardized and purchased with leverage to lower costs. Think back to the title of this book: *Zero to Hero: How to jumpstart your reliability journey given today's business challenges*. My goal in writing this book is to get readers to start their journey by helping them navigate the countless obstacles that prevent the journey from launching. Beginning with a waste focus to secure early wins, then shifting to a traditional deployment to drive long-term systems, can be a very effective "one-two punch" intervention at a plant. My process has the same ultimate goal of implementing best practice systems. I just change the order and complexity in the early months. While my model is not complex enough for many reliability purists, it works extremely well in factory cultures. I do contend that a focus on waste elimination at your site using the A3 process can be a very effective long-term reliability strategy. Consider the following:

You are on the factory floor, and you encounter a mechanic. He tells you that he does not understand why you created the kitter/stager position when there is such a large backlog. "We should be hiring more people, and you took one more mechanic off his tools." What is your response?

Option 1: *We have 31 elements to our reliability plan. We began with governance last year by forming a reliability lead team. Within 2 weeks, we crafted a vision, mission statement, and group norms. As you know, governance is one of the foundational tools of a reliability program. We then established focus teams in the following areas: work control, planning and scheduling, planned work execution, PM and PdM, and problem solving. All these teams were trained and have been working together for 6 months. The kitter/stager change came out of the planned work execution team.*

(Imagine this goes on for 20 more minutes.)

Option 2: *We are attacking waste. Our wrench time is 12%. Not because you guys aren't trying, but rather we are not giving you all the parts, tools, and equipment to do your job efficiently. We found this by walking a mile in your shoes – observing a typical day. Remember last week when you couldn't find the correct coupling? I believe this kitter/stager role will shave hours off of planned jobs and reduce our backlog. Let's give the role 90 days to see if it helps. I'll check back in with you.* (End of discussion.)

Which response is more convincing to the mechanic? Which one can they repeat to their peers in the lunchroom with a positive endorsement? No question – the simple message of eliminating waste resonates up and down the organization. Your craftsmen chose their profession because they like things to run well, they like to fix things, and they want to improve. A waste focus plays to these interests perfectly. The reliability community has turned reliability deployments into complex 40-part systems that interlink, rather than an inspiring simple message the factory floor employees will get behind. Not that employees are stupid or can't understand the complex approach, but we've pumped up the complexity so far to impress the intellectuals that we forgot the mission: "Create a reliability culture." This means all plant employees need to align. Simplicity of message is critical to understanding, and the alignment we seek.

There is plenty of room for all consultants. There is no shortage of poor reliability practices in the world. I'm offering an alternative to traditional deployments. I believe you have 4 options:

1. A traditional reliability deployment.
2. A lean driven reliability deployment.
3. Start with lean, then transition to a traditional deployment.
4. Do nothing.

My strong recommendation is to pick Option 2 or 3. Choose wisely.

Chapter 12
Selling Reliability

Selling reliability is critical. No one likes sales. Next to public speaking, I'll argue the number two fear of humans is knocking on the door of someone's house to sell them a new vacuum cleaner. But sales is a fundamental component of business, and great success comes to those that master it. Great products just sit on the storeroom shelves without sales. The same goes with reliability. I contend that reliability is an exceptional leverage point for nearly all factories due to the impact reliability has on output, cost, quality, attrition, safety, and employee morale. One lever with 6 results. Name another action that has this kind of sweeping impact. But reliability does not sell itself. There are too many distractions in our lives to just assume the decision-makers and employees know the impact reliability is having in the plant. As stated before, you cannot expect people to know about events that don't happen. If, through great PdM and problem solving, you change the mean time between failure (MTBF) of a motor from 12 months to 12 years, you need to highlight this fact. On the contrary, how fast will top management know of a motor failure costing 24 hours of unplanned downtime, customer delivery misses, and $50,000? About 2 seconds, right?

Education of reliability business impacts is best when it comes repeatedly and from many sources. Here are a few I like:

- **Weekly Email Blast**: Assign someone, or rotate the responsibility, to craft one or two emails a week highlighting a reliability success. I like to include a picture along with a short description of the success. In the description, include a historical perspective of the cost of a failure along with the new change or finding. Keep it short. Best to be just one page. Come up with a standard format. Also, don't send as an email attachment; many recipients will not open the document. Cut and paste the document into the body of the email so the picture pops up when the email is opened. This sounds trivial, but it works.

- Monthly Newsletter: Assign someone, or rotate the responsibility, to publish a monthly newsletter. Keep this simple – three pages maximum. Invite the organization to submit articles. You would be surprised at the input you will receive from some closet authors. Ensure you have a standard report section that is consistent month to month where you report KPIs, financials, what to look forward to, and help needed. Support this section with 2-3 feature articles. Include pictures of failures – especially thermal and vibration signatures, if applicable. Some organizations really get into creating a world-class newsletter. I caution you against this practice. The worst thing you can do is make the process of creating the newsletter so time-consuming that you sacrifice your reliability work on the factory floor.

- Quarterly live meeting with the lead team: Once a quarter, reliability engineers, problem solvers, OEE teams, and PdM technicians need to present to the lead team live. At my plant, this took about 2 hours and was one of the highlights of my job. Have teams and individuals report on their accomplishments, challenges, what's next, and any help needed.

- Leader standard work for meetings: With reliability being a core part of your manufacturing strategy, it needs to be mentioned in nearly every plant meeting. At my plant, we began all meetings with EHS. Find the right time, depending on the meeting purpose, to give a positive endorsement for reliability; real examples are best. Since reliability touches every facet of the business, this should not be difficult.

- "Go and see" standard work: Look for opportunities to turn a conference room meeting into a field trip. For example, in the quarterly live review just mentioned, reserve the last 30 minutes to go and see a failure or a new solution in process. Create a standard to go and see failures in the field after each morning meeting. Go and see planned work one day a week after the morning meeting. Have OEE meeting agendas include a "go and see" event during the last 15 minutes of the meeting.

- Sponsor education on PdM tools: As a PdM technicians, as stated earlier, part of your job is to take the mystery out of the work you do. Push leaders to join you for an hour on a route and have them use the tools. Bring them back to your computer for any analysis. If you find an anomaly on a route, go to the office and grab a leader. Tell them you need 20 minutes to show them something. If you solve a problem to root cause, again, grab a leader and show them the change. Sustaining sponsors takes focused effort.

One question I am surprised to get is how to calculate reliability savings. Reliability professionals become very shy when asked to provide a number for dollars saved with reliability actions. If they eventually give you a number, it is surrounded by many qualifiers to dampen the impact and offer an escape route if the full savings do not materialize. Further, they rarely want to include the cost avoidance in future budget forecasting. As a former plant manager, I need to tell you that the case for reliability savings is far more creditable than the plans your peers in quality, safety, environmental, and production are selling. These groups are selling plans that have a 40-70% chance of success, while the reliability community is reluctant to claim a change based on science with a 90% probability of success.

Why is this? Here's my thinking: it comes down to the psyche of the different disciplines. Quality leaders are very comfortable with risks. They put their stamp of approval on perhaps millions of parts per day. These people have to be optimistic to survive the stress. I believe the same goes for safety; if you stop and think about all the ways someone can get killed at your plant, it will cause you a mental breakdown. The maintenance team members tend to be perfectionists. When you balance a shaft, it is within tolerance every time. Not just some of the time – every time, no exceptions. The behavior is a characteristic of nearly everyone in maintenance; and this trait is highly prized for precision maintenance. Further, the maintenance team rarely receives positive feedback. Most of the time, they only get noticed when things break. No one calls their local electrical utility thanking them for providing reliable power yesterday; it's expected to be 100%. But if you have electricity 500 days in a row without interruption, and then the power goes out, you want to know who to call and complain. This conditions maintenance professionals to be pessimistic or have a "what can go wrong?" mindset. They are on the defensive and prefer to stick with the status quo. It's okay to add new technology like PdM, but we're going to keep doing our PMs to be sure.

For the good of the company, your plant, your community, and your family, reliability improvements that have 90% plus odds of delivering bottom-line results need to be bellowed from the tallest tower. Reliability investments depend on your selling.

Here are 3 examples of savings calculations:

1. Electrical motors – The historical spend is $2MM/year in maintenance expense. Our KPIs, vendor feedback, craft feedback, and our own observations reveal that we have poor lube practices and contamination issues (carbon or alumina dust entering motors). Poor lube is a key root cause on 80% of failures. Contamination is a key root cause on 40% of failures. (Note: It is common to see more than one root cause.)

    i. The lead team has made motors 1 of 3 focus areas for 2021. The challenge is to cut spending by 10% ($200K) through the following actions:

        a. Ultra-sonic greasing to drive precision

        b. Junction box sealing to prevent contamination
        c. Air filtration project to prevent contamination
        Note: These are science-based actions (not hopes and dreams).
    ii. We believe these actions will result in a 20% reduction in spend, but we want to under promise/over deliver, so we provided the 10% result to the controller.
    iii. Why just 10%? $200K is enough of a prize to get the attention and support of management.

2. Wrench time – Chalk circle observations have determined our wrench time to be 15%. The 2 actions below are expected to increase our efficiency to 25%. The math from observation data says 30%, so everyone felt very comfortable committing to 25%.

   Actions:

       i. Adding a kitter/stager – From existing headcount, even though backlog is growing!
       ii. Change expectations of planners to "anticipate" wrench time detractors; management should drop focus on the schedule compliance KPI. Wrench time is the goal for this year. Train the planners on best practices and require quarterly WT audits by planners.
       iii. Savings from the 10-percentage point increase in wrench time is calculated to be $500K/year. Savings will be realized through reductions in contractor spend, not replacing attrition and reduced overtime.

3. Circulating water pumps – our historical spend averages $60K/failure with 2 failures a year, or a total annual spend of $120K. We have taken the following actions:

i. Implemented PdM routes for lube and vibration. We recently had a success where we found an anomaly. We planned, kitted, and staged the job and repaired for $3,000.
ii. We commit to the team to prevent one failure per year for a net savings of $57K per year. There is a good chance we will be able to prevent all catastrophic failures, netting an impact of $114K per year.

These are all meaningful commitments based on hard science, folks. Your peers in quality, production, and safety are selling long shots. Proven reliability practices are not a leap of faith, but rather a leap of science.

Stand up and sell – your plant manager needs you.

Chapter 13
Mistakes

I've spent 12 chapters being your guide for actions to take on your reliability journey. This chapter is dedicated to mistakes, errors, and traps along your way. Many of these mistakes I personally made. All mistakes listed here I have seen, counter-measured, and coached others through. The mistakes highlighted here cripple and often kill a lot of programs. At a minimum, they will cost you time, energy, and money. Heading off on a reliability journey with the best practices in your backpack can make you confident; however, you will encounter difficulties. How you deal with these determines your fate. I have broken this chapter down into five sections:

- Excuses for delaying the start of your journey
- KPI lies you believe
- My top 10 mistakes
- Starting and stopping
- Technology is the answer.

Excuses for delaying the start of your journey

1. "Once we get caught up on our unplanned work, we will begin doing more planned proactive work." This is the most common explanation given by maintenance managers and plant managers for pushing off the start date of a reliability journey. To the experienced reliability guide, this is ridiculous. To put it bluntly, this is a statement of ignorance. These words are a giant flag waving over your plant that says, "I don't understand how reliability works." Unplanned work not eliminated by planned work or problem solving only grows. A basic understanding of reliability reveals the only method to reduce unplanned work is

planned work. The planned work conservatively is 7 times more efficient. This is the leverage you need to turn your plant around. One hour of planned work eliminates 7 hours of unplanned work in the future. This is a very conservative estimate of the impact of proactive maintenance. In my industry, it was common to see final failures cost 20-30 times the investment of proactive maintenance.

2. "We're too busy." This is related to number one above. Upon brief observation, nearly every time I discover the plant is too busy doing unplanned work to do planned work. You need to find a way to begin proactive work. This can just be one person of your team. Start with 5% planned work and claw your way up from there. You always make time for what is important. What program are you working on that has more impact on your operations than reliability? Reliability touches everything. If you are busy, you need the leverage reliability provides. With reliability you get improved quality, improved safety, more engaged employees, less unplanned chaos, and more time. Action: take a look at your calendar for last week. Calculate the hours you invested in strategic solutions verses tactical and administrative tasks. Will this percent repeated weekly drive meaningful change in six months? The number one thing you can do to improve reliability is to stop making it worse.

3. "We're different; traditional reliability tools won't work here. We've tried before." Lots of things come to my mind here. Most likely you have attempted to bring new procedures and practices into your plant through some consultant training. You then "hoped," but your culture "ate your plans for lunch." What you didn't do was a thorough current state analysis with chalk circle observation. Seeing the waste leads to a vision of target state and actions to create this new state. When a carpenter builds a table, he selects the tool to shape the wood towards his vision of the table. The vision comes first. Imagine if the carpenter just started sawing wood with no end in mind. His skill of sawing needs to

be focused on getting the final table shape. The skill of sawing is of no value without the vision.

4. "We need to hire people first. We're understaffed." I hope the story of CoatCo helped with this mistake. If you are a poor reliability plant, there is an excellent chance your current resources are grossly underutilized. Unplanned work is a black hole sucking everything into its vortex. Focus a small group on planned work, and watch your resources grow.

5. "We need dollars to invest first." For a traditional deployment, yes; for a lean driven reliability deployment, no. If the order and focus of your deployment is on waste elimination, you will get rapid results. At this time, seek a partial investment of your savings to deploy more tools and on a deeper level. Investment first was required in the 1980s; not today with our understanding of lean.

6. "We are already good. Look at our KPIs and self-assessment and tools we have deployed." Congratulations. You don't mind if I come to your plant for a week to do chalk circle observations on your results, do you? Good chance you have a lot of things going right, and a culture that embraces reliability. But there is a reason they call it a journey; you never arrive at a totally waste-free plant. The simple fact of stating that you are good and don't need help is a sign of a problem.

"Our KPIs are all good; ergo, we have achieved reliability Nirvana." KPIs are the biggest lies organizations tell themselves. I have a whole section dedicated to KPI lies later. KPIs must be supplemented with intense observation to be of any value.

Self-assessments are great for telling you the components of a reliability program you need to employ. They are terrible for scoring you on how well you have deployed them. Again, intense observation of the results of best practices and tools is what determines greatness or mediocrity. I've lived this myself. One plant I was at had the second highest externally verified audit score ever recorded by the consultant – we were "world class." We had all the tools and processes in place, but waste was rampant. It was not until we focused the tools on the waste that we excelled.

7. "Wrench time is not a problem at our plant." I wish I had a dollar for every time I've heard this. Most plants either "assume" their wrench time is good because they do best practice planning and scheduling, or they have performed some form of "snapshot" evaluation of wrench time in the field. Snapshot assessments are very common. The gist is, you effectively walk up to a job and take a virtual "snapshot" of the crew. From this view, you determine who was performing effective work and who was doing other things (like getting instruction, waiting, on break, getting tools, traveling, etc). The process is not flawed by design but is nearly always flawed in execution. Getting a representative sample is the key. The practitioner must get samples from all shifts, all times during the shift, and all crews. I have performed chalk circle wrench times on at least 15 locations; most of which were by teams of observers. These observations began before the shift and ended at the crew's end of shift. Observations were on hand-picked jobs that were expected to be the best of the best for the site. The chalk circle wrench time results were at best 50% of the plant's snapshot results. If you think your wrench time is 40%, and it is actually less than 20%, does management make different decisions? Absolutely. By the way, if the most efficient jobs were set up to be observed via chalk circle, and these yielded numbers 50% of the snapshot method, a random sampling of the plant's wrench time would result in much lower numbers. Consequently, if I'm provided a wrench time number as a plant KPI, I assume the true value is closer to 30% of the figure provided. As always, this must be verified via chalk circle observation by trained observers.

8. "We do not have top management sponsorship to start a reliability journey." Waste elimination work does not need

sponsorship. It is helpful to have, but it is not a requirement. Most managers use this excuse when they want upfront resources in the form of more planners, supervisors, and crafts because they feel they are understaffed. As you have seen from the CoatCo story, there is a good chance you have plenty of resources that are just inefficiently utilized.

Funding may be needed to start a comprehensive lubrication program, but you can get a lot of savings from doing lubrication basics with readily available information on YouTube. In my experience, 80% of the opportunity is from doing these basics. After you capture savings, seek funds to capture the last 20% of the opportunity.

9. "We are waiting for corporate to set a direction and priority." I have actually heard this, believe it or not. 90% of plant managers cringe at the thought of corporate coming in to tell them what to do. But then there is the political 10% that waits for words of wisdom from corporate. Pardon the sarcasm, but in my experience, corporate resources drive agendas in reliability that fit the average plant. Sounds good on the surface; however, the problem is, no plant is average. One plant may have a great material storeroom process, while another has a terrible process. The corporate mandate is for both to adhere to a new standard. This is good for one of these plants and a complete waste of time for the second. If, however, the corporate plan consisted of performing A3s at plants with observation as the hallmark of the current state evaluation, I would wait for corporate. However, I only know of one company that operated in this manner, and the director of their program retired. You may know him; he is the author of this book.

10. "We're still improving our plans and learning the skills to be successful." Don't be paralyzed by waiting on the perfect plan. Whatever plan you complete will be wrong. The plan will encounter unforeseen obstacles and you will need to adapt. The trick to starting a reliability journey is to take the first step. As a

rule of thumb, if you have waited over 30 days, you are now procrastinating. Jump in. Learn by doing.

KPI lies you believe

1. PM Compliance – Your plant is at 90% PM compliance; you're pretty good, right? However, under evaluation you discover that several critical production centers have missed PMs for several months. The high results were due to completing 100% of the more numerous non-critical PMs. Example: The PM inspection of the lighting in the breakroom has the same impact on the PM compliance metric as completing the PM on your 5,000 horsepower reversing mill motor. Secondly, does PM compliance tell you anything about the level of precision and completeness of the PM? The answer is no. While it is rare for an electrician to cut corners, have you ever had your planned outage cut from 8 hours to 2? How do you think this was accomplished? They were told to cut corners. Worse – 10 times out of 10, the PM was closed as complete boosting the metric value.

   The remedy? Audit. Leadership must create standard work to audit PM execution as part of their weekly schedule. This takes place in the office and field. In the office, ask to see the 10% that did not get executed. On the shop floor, randomly select PMs that were executed last week and look for physical evidence of the work being performed. Secondly, talk to the craftsmen who performed the work. Discuss the job findings and see if the discussion matches the notes on the workorder. At one of my plants, it was common to see a work order with 20 checkmarks and a signature on it. When discussing the job with the mechanic, he would bend my ear for 30 minutes with issues related to the equipment. Something is wrong here, gang. In my example, a true story by the way, the mechanic gave up caring about the equipment, because no matter how much he wrote down, no follow-up action would be taken. His actual quote: "You stopped caring, so I stopped caring."

2. Schedule Compliance – Most planners are measured by schedule compliance. Over a period of years, it is common to find a job that was originally scheduled for 2 hours to have evolved into an 8-hour job. Why? Because 5 years ago, during a PM, we incurred some discovery work that took the whole shift to complete. This resulted in many jobs being missed on the outage, and an overall low schedule compliance metric. The whole team, and especially the planner, was scolded by upper management. The metric had turned "red" on the dashboard that was seen at corporate. The planner was not going to let that happen again. He changed the PM to 8 hours. If the work was completed in just 2 hours, the crews could be assigned more work (in theory, not reality). Now repeat this story every day for 25 years. How precise is your schedule?

The remedy, as always, is verification through observation. Poor precision in the schedule can be easily seen in wrench time. Secondly, the morning after an outage, when maintenance is getting a pat of the back for completing an outage early, you may want to do 5 whys problem solving.

3. Schedule Efficiency – I have never found this metric to be accurate or useful. This metric compares the planned job time with the actual job time. No one ever self-reports they scheduled too much time for a job – for you never really know how long the job is going to take next time. If there are excess hours assigned to a job, everybody is happy. The planner gets the work accomplished with all KPIs at 100%; the supervisor is happy since the work was completed, and she had time to do some other work that was not on the schedule. The crew is happy because the work was completed, and they had plenty of time to execute it. Everyone is happy, except the person paying the bill. Unfortunately, the owner is in many cases a corporation. Imagine a contractor gave you a bid to build a fire pit in your back yard. The bid was for $1,000 and included 20 hours of labor. How

happy would you be if the work only required 10 hours of labor, and you still got the bill for $1,000?

4. Percent PdM work hours – The largest error I see with this metric is having supervision pull these resources into reactive maintenance or into planned outage work. Most often a PdM team's hours are charged to standing (open) work orders and 40 hours are charged to PdM every week with no regards to reality. This can boost this metric as much as 50%. The remedy is quite simple: have each PdM technician submit their hours each week to the plant's leadership team. If you do not have a lead team, supply the hours to the plant manager. Secondly, require approval from the maintenance manager to pull these resources off of PdM. I've found just by implementing this approval, the hours pulled goes down by 90%. The PdM team is just too easy to pull out of their long-term strategic work. Plus, they are often the most highly motivated and skilled members of the team. There are other options for supervisors: prioritize work bumping some non-urgent task out 24 hours; asking for overtime; or securing contractors to do the work. What everyone on the team must realize is that finding by a PdM tech is saving 7-20 hours of breakdown maintenance in the future. Recall that overtime is paid at 1.5 times base pay. While expensive, this is far less than the minimum 7-to-1 impact you get from proactive work. This is a difficult concept to comprehend in a 100% reactive culture.

5. Planned work percent – I've seen a lot of creative ways to show inflated planned maintenance numbers. One is to assign all crafts planned work for 8 hours a day, 5 days a week. Require them to address all unplanned work within this assigned time. In doing so, you are at 100% planned work. But as stated before, the planner adds "slop" time into the planned jobs to account for being pulled into unplanned work. Another error, as stated previously, is to pull PdM technicians into unplanned work, leaving their hours as planned PdM. I've even had plans that call unplanned work planned work. "The planned work for Joe next week is to do unplanned work; therefore, the work is planned."

This is rare – but it proves you need to audit what is happening. One reason planned work percent gets inflated is due to management expectations. A vice-president goes to a reliability seminar and learns the value of planned work. He comes back to the plant and issues an email that states all his plants are to report planned work percent, and all need to provide reasons why they are not at 85% efficacy Monday. Have you ever received this email? How many plants have had their planned work go up rapidly without any real changes on the shop floor? I've found it to be fairly common.

My top 10 mistakes

Below I rank my top lessons learned. These are listed in order of impact that they had on me and my journey. Most of these I have already introduced in this book; two are new. Following this list, I will detail these new mistakes further.

1. Not using observation to supplement my data and direct my reliability best practice actions. Establishing a "go and see" culture.
2. Not having an overall waste focus in my reliability focus. Too much attention was given to the understanding and application of tools alone, and not the results I wanted to achieve on the factory floor.
3. Not understanding that I needed to address emergency work first. Uncontrolled, it will be a massive tsunami to all your plans, wiping them out.
4. I assumed sponsors and top management would know my results; I did not have to sell reliability.
5. Problem solving is just another best practice tool to deploy. In reality, it is the key to dramatic results.
6. Believing KPIs and metrics were accurate.
7. Blaming the maintenance team's work ethic for our reliability problems.
8. Part-time reliability engineers can work. This is an illusion.

9. I will not need to make personnel changes as part of creating a reliability culture.
10. I don't need a coach.

One of these mistakes that I've yet to cover is #7: Blaming the maintenance team's work ethic for our reliability problems.

I will explain this mistake by example: Superficial observations at my plant reveal that my employees start work late, take long breaks, and quit early. Total opportunity to improve is 2 hours. I then have a supervisor meeting where I firmly tell them to manage abuse of breaks, lunches, shift start, and shift end. It is a minimum expectation to manage these. By Monday, the problem appears to be gone. I send an email to the supervisors telling them "good job."

The question for you is: Did the two hours of waste get converted into productive wrench time? "Heck no" is the answer; employees just found another place to be other than in the breakroom where management can see them. Here's what was going on in reality: The crews did not have enough planned work assigned to them. They were given 3 hours of work to do in 8 hours. They got it done. The problem is the same as the issues we had earlier in this book on wrench time. Nearly every plant I have audited via chalk circle had "no work assigned" be the number one detractor of wrench time. At the same time, the backlog was growing and the plant was working 20% overtime.

I have never audited a plant and found the craft team's work ethic to be lacking. I always found huge system issues that prevented employees from excelling. Consequently, I'd be very careful jumping to conclusions about work ethic. The odds are, you are incorrect and you will lose enormous respect and creditability going down this path. Now, I am a realist – everyone reading this book could give me a handful of poor-performing employees; however, this is less than 5% of the population. Make decisions based on the 95% that want to come in the plant every day and do a good job. It's empowering and respects the 95%.

The next new problem is #9: I will not need to make personnel changes as part of creating a reliability culture.

If your plant is over 20 years old, you have had people excel in the established culture. Some people may be great at reactive work. They love coming into work and "saving the day" with their heroics of rapid problem solving and innovative solutions in the moment. There are also planners that took their roles because they were tired of being supervisors. They do the minimum ordering of parts and schedule the work for 8 hours, doubling the time they think the job could take. This keeps their metrics high. They put in their time to earn this "easy job." There is also the maintenance manager that refuses to accept new processes, loves his time-based maintenance, and sees only risks with change. All these people will struggle in a proactive best practice culture dedicated to aggressively attacking waste.

We must give everyone the time, training, and motivation to adapt to the new culture we are trying to create. However, there will be people that don't make it. How you handle these people will be observed by the entire organization. Some may need another assignment to excel, and some may need to exit. What you cannot allow to happen is for them to stay and resist. In my personal work experience, I had to make personnel changes as part of every change effort.

Starting and stopping

I'd estimate that 90% of the plants that have started a reliability journey have been asked to stop or pause progress due to other business priorities. Most often, that priority is a negative business cycle where spending needs to be carefully controlled and scrutinized. Asking the lead team to pause a reliability journey is a clear signal the leader has no idea what reliability is all about. Yet the practice is very common. Most times, I believe this action is due to ignorance, but for some, reliability gets wrapped up in the directive to "cancel everything non-essential." This is why it is critical to understand and sell reliability as a list of best practices that attack waste at a plant. Waste reduction is exactly what you should be accelerating in difficult economic times. The time to sell reliability is always. Further, this is exactly why you need a deployment that is not front-loaded with cost and delivers a prize 3-5 years down the road. By employing lean concepts, investment requests come after results.

It is also critical to note that stopping, or even pausing, a reliability journey is costly and has long-term cultural impacts. It is clear to most that without proactive maintenance, assets will degrade over time, resulting in lower and lower reliability, leading to poorer and poorer financials. A pause will also result in a massive work backlog that must be overcome to reenter your reliability journey at the same place you stopped.

As an example, a 6-month pause can result in a 12-month recovery to get to the same point of asset, process, and system health. What is less clear to leaders is the emotional investment the operators and maintenance teams put into the reliability journey. Stopping the journey drives cynicism and distrust. It is illogical to assume a 6-month pause can restart as if nothing changed. It is more typical to see a 6-month pause result in a setback of one year. Some reliability "champions" may never recover from this lack of trust with management. This is the fundamental problem with a lengthy deployment.

Technology is the answer

There are many plants that have deployed best practices, yet exceptional results elude them. Frequently, managers will reach for technology to propel them to great financial outcomes. Some examples include automatically generated dashboard reports from CMMS data, available on your smartphone; paying an outside contractor to perform machine learning and artificial intelligence on CMMS data; and adding remote condition monitoring and analysis to your critical assets. All of these are great services if placed on top of a sound foundation. However, scoring well in a traditional best practice audit does not equate to a solid foundation.

Two massive holes I've seen with this logic:

1. The best practices do not address the waste in your culture.

2. Problem solving is lacking.

Just by telling me your plant is doing great with best practices, yet missing results, you are telling me you are not focusing on waste. If I can kick a soccer ball 40 yards, that does not mean I can pass the soccer ball to another player in stride to kick a goal.

The CoatCo Current State revealed a completely different picture of reality than we first understood from the Business Case. Technology tools must be applied as a solution to an observed problem, and not a solution in search of a problem. As an example, let's look at the now-popular KPI dashboards on smartphones from CMMS data. Two problems with this approach:

1. There is an assumption that the data in the CMMS and the resultant KPIs are accurate. This is NEVER the case for an emerging reliability system, and rare in mature systems.

2. Top management will begin micromanaging the plant on KPIs where the status aggravates them. For example: Planned work percent is below the standard of 85%, turning the KPI red. Management tells the plant to give them a plan to make it green in 2 weeks. Magically, the metric turns green. I've seen this happen for 2 reasons: one, the plant dropped higher priority actions to change the status of this metric, and two, the plant changed the definition of planned work

to artificially change the status color of the metric. The first of these will drive reliability down; the second accomplishes nothing.

Consequently, the majority of the plants at an emerging point on their reliability journey will receive micromanagement of their actions based on bad data. This will create frustration by all involved, along with worsening results. Great tool – wrong application.

*****

I have covered a lot of topics, processes and lessons learned in this book. I hope you gleaned at least one new tool for application along your reliability journey. The big takeaway I hope you employ going forward is to "go and see". Thoroughly understanding current state reveals simple and often "no cost" solutions for rapid results. In my 33 professional years, I've never had a more powerful tool for culture change than chalk circle observation. Win every debate by adding observation to your data set. Gain organizational alignment and enthusiasm through a mantra of "a relentless focus on waste".

Good luck.

Quick Reference Guide

1. Core Beliefs of Lean Driven Reliability

2. Format of an A3

3. 7 Forms of Waste

4. Rules in Use

5. Chalk Circle Observation

6. ADKAR

7. Warren Buffet's 5/25 Rule

8. Listing of Common Current State Wastes and Potential Actions

Core Beliefs of Lean Driven Reliability

> A relentless focus on seeing, understanding, and eliminating waste will rapidly deliver a reliability culture at the lowest possible cost.

1. By combining lean concepts with reliability best practices, rapid, sustainable, and dramatic results can begin in just weeks. The dogma of required upfront investment with bow wave spending for the first 3-5 years is a choice, not a rite of passage.

2. Every reliability tool and best practice is targeting waste.

3. A culture of "go and see" is fundamental to knowing plant wastes and the reality of how the plant works. Leaders must be trained to see the waste in their plant through chalk circle observation and the 7 forms of waste.

4. It is insane to implement new tools and best practices without intimately knowing the waste you want to eliminate at your plant.

5. A waste focus is not the norm. Nearly all training, trade magazine articles, books, and seminars are focused on tools and best practice training to purchase – all great stuff in due time, but the change agent must know your order of deployment.

6. KPIs can be very misleading and lead to many bad management decisions. Assume they are wrong until you verify all assumptions and validate them with chalk circle observations.

7. 75% of waste does not require the plant to buy anything to eliminate/reduce this waste.

8. Maintenance leaders need to take risks to improve current state and provide hope for the future – risks that include stopping historical, traditional practices and accepting best practices.

9. A resourced problem-solving system is required for dramatic results.

10. Your maintenance team will love your focus on waste, for it makes their daily life more predictable and fulfilling. They too like to see equipment running reliably.

Suggested vision for your plant...

Relentless focus on seeing, understanding, and eliminating waste.

Simple, easy to repeat, and hard to disagree with the concept.

Format of an A3

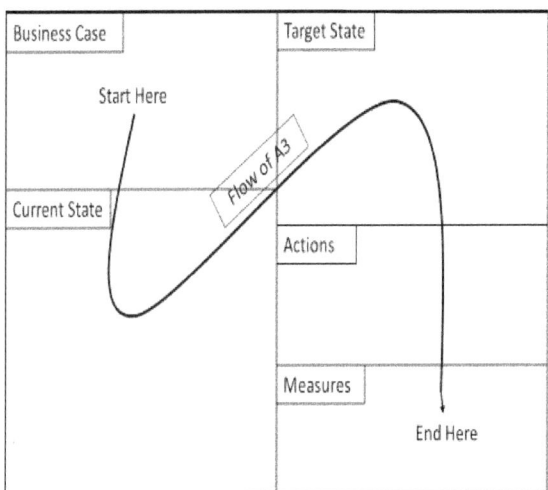

In the early years of using the A3, you were taught to always handwrite the process with a pencil. Pictures were preferred over text (imagine stick figures), for if you can draw it, you understand the concepts to a deeper level. Today, nearly all A3s are computerized, yet most still contain pictures.

Checklist of A3 Steps

1. Develop the Business Case with plant's leadership team.
2. Verify Business Case with peer one-on-one floor meetings.
3. Find Current State through chalk circle observations.
4. Detail your Target State with the plant lead team.
5. Craft Actions required to achieve the Target State
6. Select the Measures for success. Focus on a few.

7 Forms of Waste (TIM WOOD)

1. Transportation Waste – Moving product, materials, tools, or equipment from one place to another on the shop floor. In many cases, this movement is required, but ask yourself if the customer is willing to pay for it. Regardless, it needs to be reduced as much as possible. Example: A mechanic walking back to the shop to get a wrench. Yes, the mechanic needed the wrench, but why did she have to make a second trip?

2. Inventory Waste – Having money tied up in excess inventory.
   Example: Having five spare motors in stores when your annual consumption is one.

3. Motion Waste – Non-value-added movement by employees. Example: A mechanic installed the wrong pump. Upon realizing the mistake, he found the correct pump and installed it. The error delayed the job 3 hours.

4. Waiting Waste – Waiting for instruction, parts, processing, or help from others.
   Example: Two mechanics waiting for a part to be delivered from the storeroom.

5. Over-Production Waste – Making too many parts.
   Example: Maintenance built 10 molten metal pump systems as spares when monthly consumption is one.

6. Over-Processing Waste – Performing more work on a process or material than is necessary to meet a standard. Example: A plant has a lube sampling program as part of its predictive maintenance plan. If the oil is changed monthly on a PM, the PM would be considered over-processing. It is a step that is not required to meet a standard.

7. <u>Defects Waste</u> – Non-conforming material. Material does not meet the standard.
Example: A new motor was installed on the crane. The motor shaft was not aligned properly, and the newly installed motor failed in two days.

Note: Recently some professionals have added an 8th waste – Skills. The waste stems from not utilizing the skillset of all employees. Example: Not using input from operators and maintenance personnel during problem solving. I like this new waste but chose to add as a footnote rather than modify the list. To remember the 8 forms of waste, just add an "S" to TIM WOOD – TIM WOODS.

Rules-in-Use

In a paper titled "Decoding the DNA of the Toyota Production System," Steven Spear and H. Kent Bowen (*Harvard Business Review*, September-October 1999 issue) attempted to break down the culture that exists in a lean plant. They simplified lean to four rules:

1. The Work of One
2. Connections
3. Fixed Flow
4. Improvement

I will further simplify and apply these concepts to reliability and maintenance.

The Work of One – The tasks of each worker are highly specified for content, sequence of steps, time required, and outcome.

> Example: If a mechanic is to replace a fan belt, does he know all the steps in sequence to perform within a specific time window? Also, how is the work tested to ensure the repairs do not contain defects? A good job plan provides this information.

Connections – Rules or norms exist between workers on how they work together or how work is transferred from one to another.

> Example: After the electrician locks, tags and verifies isolation of the overhead crane, he radios the mechanic to begin his work (after he validates tags and puts his lock on the lock box). This has proven to be a major source of waste in my experience; especially connections across shift changes.

Fixed Flow – There is one right way to do work.

> Example: Work orders are submitted electronically to the planner by the requester. The planner does not accept phone calls, meeting comments, nor handwritten notes as work orders.

Improvement –There is a system to drive improvements, and everyone knows it.

> Example: We have bi-weekly OEE meetings for our production center. Operators, crafts, engineers, planners, supervisors, and technical assistants are all invited to participate.

Chalk Circle Observations

Taiichi Ohno of Toyota fame defined the term "chalk circle" observation. Chalk circle refers to an imaginary circle drawn on the factory floor in which the observer is confined by for at least four hours. Eight hours and multi-day observations are preferred, as well as multiple observers (number of observers depends on plant size). The target of the observation is to find the existence of one or more of the 7 forms of waste. This is critical data for the creation of current state when using the A3 problem-solving tool.

Many leaders observe for a couple of minutes on their way from one meeting to the next; from this, they draw conclusions based on this snapshot of reality, KPIs, and opinions expressed in meetings. The premise of chalk circle is that you really cannot grasp the intricacies of a process in just minutes – much less from KPIs. While doing chalk circle, it is common for me to think I'm wasting my time, but then after two hours – bam! There it is – a gold nugget of waste I would have never imagined. The vast majority of leaders spend no time in chalk circle, and thus are doomed to make bad decisions based on an incorrect assessment of reality. The irony is that if you interview leaders, they will tell you they are a "floor leader." This is the disease impacting manufacturing today – leaders are making million-dollar decisions based on biased opinions, filtered electronic data, and an illusion that they have observed reality.

Culture change tool: ADKAR

ADKAR, by Jeff Hiatt, is a popular culture change model. The model name is an acronym: Awareness, Desire, Knowledge, Ability and Reinforcement. Each of these components must be addressed for successful change.

- Awareness of the need to change. Why should we change from the current way of doing things?
- Desire to support the change. The organization is convinced we need to change.
- Knowledge of how to change. What changes am I supposed to make?
- Ability. I have the skills and training to change to the new desired behavior.
- Reinforcement. Leaders provide acknowledgment and appreciation to those that execute the change. Supervisors also provide coaching to those who need help.

Note: If communicated clearly, the A3 process does an excellent job detailing awareness, desire, knowledge, and ability. I believe the model misses reinforcement. This is the primary reason I referred to ADKAR in this book and included reinforcement as an action in the A3.

Warren Buffett's 5/25 Rule

Simply put, focus is the key to success. Warren's strategy is to prioritize the most important goals and achievements in your life and "avoid at all cost" the trivial many. He reminds us that it is not what you do that determines your success, but what you don't do. You can become a world class pianist if you focus on playing the piano. But if you dabble with the guitar, flute, drums, and trumpet, you will be competent at many but an expert in none.

Applied to reliability, a reliability lead team can accomplish great change if they pick 3-5 areas to improve at the plant. If you can "go deep" in these few and align across the organization, you will progress towards greatness. Each year, the focus can change to a new 3-5, but use caution to ensure you are not jumping to new target states without locking in hard-fought gains. Far too many plants focus on 30 KPIs and make glacial progress on them all. This behavior is encouraged by corporate, ignorant top leaders. They can play a game of "gotcha" when they find a missing component of your reliability program. Communication and getting buy-in early from corporate and senior leaders are key.

You may begin by showing this YouTube video on the power of focusing: https:/youtu.be/gkhtYs22bLI.

Listing of common current state wastes and potential actions

Here is a list of common wastes I have found in my travels, and actions I have recommended or utilized myself to reduce/eliminate. You may not be able to directly copy and paste them, but they should stimulate ideas for you.

Waste 1: Poor maintenance wrench time (under 30%)

Problem:

The vast majority of plants (and this could be 100%) grossly overestimate their wrench time. I've completed detailed wrench time studies at over 15 different plants. I've found that wrench times are overstated by at least 100%. I found many at 200%. Not only is this hiding enormous waste, it directs leadership to make bad decisions, just like in the example with CoatCo.

Imagine you have 50 mechanics, and you are at 40% wrench time. Now 10 mechanics have decided to retire this year; do you replace them? Probably so. Now let's change the wrench time to 12%. Do you replace the 10? No way; this is a $1MM opportunity. You will get far better results by improving your systems. Keep this gross over inflation tendency in wrench time numbers in mind as you evaluate your plant's wastes. As stated plenty of times, use chalk circle observation to determine your real wrench time.

Potential Action A:

The best, quickest, and lowest cost change I have made in my career is to create a kitter/stager role. Kitting is putting together every part (including shop parts and special tools) in a basket or on a skid. Your goal is to have no reason for the craft team to come back to the shop or drive around to other maintenance shops to find parts. Do not discount the value of this. Excellent wrench times begin about 50%.

An 8-hour shift would have 4 hours of productive wrench time. Consequently, the loss of just one-hour results in a 25% wrench time drop! Going back to the shop to get a bolt can easily cost a job this hour. How? Consider this: The shop does not have the right bolt, so the mechanic drives to another shop on site, where he runs into a friend he has not seen in weeks. They have a short discussion. Now it's getting close to breaktime. He takes his break. Next, his supervisor asks him for some help on an unplanned job, because he sees him in the shop. This may sound made-up, but I've observed it.

Staging includes placing the kit and all necessary equipment (mobile cranes, lifts, welders, etc) within 10 feet of the job execution site. Have production designate a permanent staging spot for you by painting a yellow box on the floor. Staging too can boost wrench time another 50% by eliminating the time searching for the kit, a fork truck, a welder, or aerial lift.

This role is to be funded from existing resources. I have seen wrench times improved from 15% to 30% almost overnight with this action. In doing so, your workforce for planned work just doubled. Do not wait for a good time to make this change. Do it now. If you have 30 craftsmen, and your backlog is growing by the day, this is the perfect action to take. Would you rather have 30 mechanics at 15% wrench time, or 29 mechanics at 30% wrench time? It's not even close, but let's do the math:

- 30 mechanics x 40 hours/week x 15% wrench time = 180 hours/week effective work
- 29 mechanics x 40 hours/week x 30% wrench time = 348 hours/week effective work

This is a 93% increase in efficiency for free. I like to have this person shift their schedule forward two hours a day to perform timely staging.

Example: If the normal crew schedule is 7am to 3pm, have the kitter/stager work 5am to 1pm. As a rule of thumb, one kitter/stager can perform effectively for about 25 planned work personnel. Also, it is common for this person to serve as the crib attendant.

Potential Action B:

Ensure your maintenance planners know that they own wrench time. As part of their planning and scheduling process, they must anticipate wrench time detractors. This should not be too time-consuming, but can save hours for crews.

- Example 1: Even though the planner ordered the exact same gearbox to be installed, the base plate, shaft diameter, and keyway should be physically measured to verify it is in fact identical.

  Example 2: If the work being planned has a complex energy isolation plan, review the process with the craft team, executing the work the week prior.

- Example 3: If the work requires operations to place equipment in a particular position to allow maintenance work, meet face-to-face with the operations supervisor the day before work is to begin to see if they have any concerns or questions. Anticipation of wrench time detractors should be in the performance expectations of planners as well as their job description.

Potential Action C:

Train planners on chalk circle observation. Require planners to perform one observation per month and document learnings in an email to the maintenance manager.

Waste 2: Our backlog is growing and we cannot complete all planned work

Problem:

PMs can make up a huge portion of your weekly planned work. This can make it very difficult to complete repair work, resulting in your backlog growing.

Potential Action:

A major catalyst at one of my plants was to conduct PM kaizens (some call these PM optimizations). I like calling them a kaizen to drive home the point of quick action. My team developed PM Kaizen I, II and III.

PM Kaizen I was designed to get immediate gains. We printed out all the PMs for a whole department and assembled all stakeholders for the department in a conference room. Include planners, supervisors, engineers, technical assistants, and crafts. We then divided up into teams. The goal was to review each PM for gross errors in resources (number of crafts), duration (how long it takes to execute the PM), frequency (how often the PM is executed) and finally, whether the failure mode on the asset was also covered by PdM. If so, we looked at cancelling the time-based PM and retaining the PdM. Your goal is to quickly capture gross errors. A good rule of thumb is one minute per PM in this kaizen. Do not optimize the content of the PM during PM Kaizen I. The whole process should take 1-2 days depending on the size of the area your team chooses to address. It is common to cut 10-20% of your PM hours in PM Kaizen I – a great investment of a couple of days. Here's a few areas of significant waste to be on the lookout for:

1. Does the PM include time for discovery work? This is very common in older plants and a very poor practice. The reason PM hours tend to grow is that planners and supervisors are judged by PM and schedule compliance. If something goes wrong in 1 out of 10 PMs, they don't want it to impact their performance numbers and have a performance discussion. The easy solution is to plan for worst case and thus inject "hidden" waste into the process.

2. All PMs are scheduled for 4 or 8 hours. This is also a result of holding planners and supervisors to KPIs that encourage them to put fat into schedules. Another root cause of duration inflation is when crews are expected to do emergency work and planned work. The planner inflates the hours to accommodate both tasks. Do you see how this fact inflates your planned work percentage and hides unplanned work? In this case, all unplanned hours are charged to planned work orders, distorting the KPIs while losing all record of the unplanned work. A good practice is to estimate the time taken to execute the PM and add 30 minutes for problems and travel time to the next job.

3. PMs always have 2 or more resources assigned. Someone once told me PMs are scheduled just like Noah's Ark: 2 by 2. Look for opportunities to drop one resource. Again, this is a rapid look for gross errors. If the team begins to debate the issue, just stop. Save the problem for PM Kaizen II.

PM Kaizen II is intended to repeat PM Kaizen I with just a little more time spent in evaluation. Rather than the one-minute evaluation, PM Kaizen II will average closer to 3 minutes each. Some will be longer, and some shorter. Short debates are acceptable. You may be thinking that this is a wasted step, but I've found that kaizen participants are reluctant to remove hours in PM Kaizen I. After completing the process once and seeing the results, they are less fearful of mistakes. Therefore, the second event can produce another 10-20% hours reduction. Consequently, I like to have PM Kaizen II at least a month after PM Kaizen I to address the fear concern. In my experience, PM Kaizen II produced a greater hour reduction that PM Kaizen I. This process will take 2-3 days depending on the scope of assets.

PM Kaizen III more resembles a Simplified Failure Modes and Effects Analysis (SFMEA). A reliability engineer needs to be present. This phase looks at failure modes and determines if the PM will address the failure mode or just make everyone feel better. Several PMs I've eliminated were what I call "feel good PMs." They have no basis in science, but they allow managers to sleep better at night. Often these develop from external quality or significant safety events. A small team may spend 30-60 minutes on a PM in this phase. Included in PM Kaizen III is an evaluation of job steps, sequence, and quality check. In my experience, this kaizen produces results similar to PM Kaizen II but will take much more time to complete. This step may take 6-8 months. You may be tempted to skip PM Kaizen I and II, but my goal is to front-load the savings. By front-loading the easy wastes, you free up resources now. This can result in you expanding PdM, problem solving, capturing attrition, funding a kitter/stager, etc.

Results:

I had one plant cut their total PM hours by 50% with this three-pronged attack. This was right at 100,000 hours. That's 50 full-time equivalent persons (FTEs) to redirect to planned work, reduce outside contractor spend, or allow to attrit out of the plant without replacement.

You will make some mistakes. Some equipment will fail after you eliminate a PM. At my plant, we made mistakes on about 1% of the PMs. You must accept you will be wrong; however, I've found the prize (50% reduction in PM hours) to be well worth the cost.

Waste 3: We do not have a problem-solving system

Problem:

Big results in reliability come from deploying best practices – one of which is problem solving. I've found problem solving to be my biggest lever to savings and culture change. It is critical that the entire organization understands how they can get their ideas and input heard through a structured problem-solving process.

Potential Action A:

I've said it before – part-time reliability engineers do not work. Many plants divide up their engineers into process departments. The engineer is expected to assist with all reactive maintenance and existing restore flow concerns, plus lead capital improvement projects, plus act as a reliability engineer. The urgent always dominates the strategic. Consider restructuring your engineering coverage to free up at least one engineer to do full-time strategic reliability improvements. This may seem simple, but you will be shocked with the results. At one of my plants, I freed up two resources by restructuring and saved $2MM in the first year. These savings compounded year over year; plus, we added new findings each year.

Potential Action B:

Begin using OEE teams. Have the lead team sponsor OEE teams on critical assets. Begin with 1-3 teams. Assign a lead team member as a sponsor of each team. This shows importance and helps with obstacles and resources.

Potential Action C:

Start a failed part autopsy area. How many times do you come into work and find out a part failed and was thrown in the dumpster (and perhaps hauled off to the dump)? An autopsy area can be a series of shelves for placing failed parts and a work table to disassemble components. By creating an autopsy area, you put in a physical change that tells the organization you will be tearing down equipment to find failure mode. Once the defect is found, advertise the failure in email and tool box meetings with the crafts. An autopsy area was critical at one of my plants as we moved from a PM culture to a PdM culture. Many of the crafts and production employees did not understand why we were changing out a perfectly good pump. Their views were changed when we showed them the bearing in the early stages of failure.

Waste 4: Unplanned work is consuming our workday

Problem:

Unplanned work is a top killer of your reliability plans. It will manage you if you don't manage it. Every plant must address unplanned work, or you will be doomed to fail.

Potential Action A:

Fact: The only way to reduce unplanned work is by performing more planned work. Managers frequently tell me that they will start their reliability journey as soon as they get caught up on their emergency work. Folks, this is insane. You must begin planned work. If you have 30 crafts in reactive maintenance, pull one to do planned maintenance. If she is working on the right stuff, unplanned work will go down. Then put 2 crafts on planned work, and so on. Best practice is 90% planned work, but you may have a short-term goal of 10%. It's a start.

Potential Action B:

Separate your planned work and unplanned work crews. The urgent always trumps the strategic. The emergency work will always be completed before the bearings are greased proactively. Many locations assign crews to an area and expect them to handle all emergency calls and planned work. How is the planner supposed to estimate time for such a crew? They have no idea how much they will be pulled off the planned work. This is one root cause for PM duration inflation, as discussed earlier. The best practice is to assign an emergency crew to handle all emergency calls and have a planned crew to execute planned work. The manager needs to closely monitor workloads and be the gatekeeper for pulling planned resources into unplanned work.

Potential Action C:

Start a PdM lite program. Now, this is going to make some PdM purists cringe, but I have had great success giving a mechanic an IR camera and telling them to "look at everything moving." Setting up PdM routes takes time. So, I created workorders that would say, "PdM lite scan the green mill" – one mechanic; two hours. The mechanic would walk the process and put the camera on anything that could generate heat while in failure mode. They looked at motors, rollers, belts, hoses, hydraulic systems, etc. They always came back with at least two findings. These findings turned into planned work rather than unplanned work. The rule of thumb is that planned work is 7 times more efficient than unplanned. However, in my plants, I found this ratio to be closer to 10, 20, and even 30 to 1. This leverage is how you win the unplanned work battle.

Waste 5: Poor management decisions

Problem: Managers are not idiots, but most draw conclusions from bad data. We have talked extensively about supplementing KPIs, electronic data, and opinion with observation. However, so far we have not talked about bad data. A good rule of thumb is to assume all CMMS data is bad. You read that correctly – assume the data is bad until you have proven otherwise.

> Example 1: The planner plans a pump replacement for 8 hours. The actual work takes about 3 hours, but the seasoned planner knows the mechanic will be pulled to do emergency work in the area. At the end of the day, all work is charged to the planned job work order. This behavior is in place in most

plants. Consequently, there's a good chance your planned work percentage is garbage. You should try to correct bad data at every opportunity.

Example 2: Your PdM percent of total work hours is an impressive 20%. However, upon investigation, you determine that your PdM technicians are being pulled into outages and unplanned work 2 days a week. Nevertheless, their hours total against a standing workorder for PdM.

Example 3: Your PM compliance last month was an impressive 90% – nice job. However, you determine that a bottleneck production center has not had a PM executed in 6 months; how can this be? Well, all PMs count the same towards this percentage. The PM on the drinking fountain has the same impact as lubricating a critical bearing on a bottleneck production center. I'm not trying to be cynical, but basing decisions solely on data and KPIs is a bad practice.

Bad data exists everywhere and with everyone. I was consulting at a casting plant where I was challenged to increase output by 20%. This casting plant made solid aluminum ingots in a batch process. In a batch process, turnaround time, or time to stop one cast and start another, is critical to throughput. The plant's average turnaround time was 55 minutes. However, in my observations, I witnessed 50, 82, 54, 49, and 91 minutes on consecutive casts. These averaged to 65.2 minutes. Hmm. Come to find out every turnaround that occurred during a shift change was discarded in the data. The plant's 55 average only represents about 60% of the data. The manager thinks she is pretty good, since an outside consulting firm stated that 50 minutes was world class. A massive opportunity exists during shift changes, and no one is working on it. Why? Bad, misleading or incomplete data.

Potential Action A:

Every leader should be extremely cautious when accepting data and KPIs as fact. You will be far better served by assuming they are bogus. For practical purposes, I do use data and KPIs, but only to point me in a direction to observe. If I use the data, I guarantee it has been validated with chalk circle observation.

Potential Action B:

Establish a "go and see culture." If I could snap my fingers and make one thing happen at plants to dramatically improve their reliability, it would be this. Leaders do not spend time on the shop floor seeing reality, and they think they do. I only count 4-hour observation events towards this metric. Here are some actions you can take to improve current state:

1. The plant manager requires all lead team members to put 4 hours on their calendar each week for observation. I suggest establishing a consistent time for each manager to enable meetings to continue. For example, pick Wednesdays from noon to 4 p.m. Note that this includes a shift change. It is best to rotate the timing of this observation monthly. If this forces meetings to change, consider what your organization values more – observation? or meetings?

2. Create standard work after the morning meeting to "go and see" recent failures in the field. Have the planner bring recent PMs and workorders for the equipment.

3. Create standard work once a week after the morning meeting to "go and see" planned work being executed.

4. Create standard work during OEE meetings for the team to "go and see" improvements in the field.

5. Require lead team members to sponsor a PdM technology and shadow the execution of a route.

Potential Action C:

Start a reliability lead team. A leadership team consisting of the production manager, maintenance manager, human relations manager, controller, engineering manager, and quality manager is a key characteristic of reliability-driven plants. The team sets the vision, mission, and the areas of focus. The lead team members serve as sponsors to OEE teams, PdM teams, and other focused efforts. Major tasks of the lead team include monitoring measures of success, providing alignment and focus to the reliability efforts, listening to help requested by teams, and having teams report regularly on progress.

Waste 6: Operations will not own reliability

Problem:

This topic was discussed in the A3 section of this book. At the majority of plants, reliability is owned by maintenance. However, this is an illusion. Operations approves all equipment outages, what work gets completed in outages, when the outages occur, how much we spend on equipment, how the equipment is used, and where resources are assigned by setting priorities. Operations leaders are clearly the "deciders" when it comes to plant reliability. Maintenance is responsible for efficient execution of maintenance, implementing best practices and problem solving. From this work, maintenance personnel provide advice to the operations leaders. Consequently, the maintenance team are "advisers" to reliability. Many consultants propose a partnership between everyone to be the best target state for reliability ownership; safety is often given as a model. I do see their point, but I contend that if everyone owns reliability, then no one does. Further, by making operations the leaders of a reliability journey they will quickly determine they need "teammates" to be successful. In a highly mature reliability culture (and this is about 1% of plants), a partnership may be the best target state.

Potential Action:

Training is the remedy; it's best if it comes from outside your organization to have increased creditability in the changes required. Secondly, accountability for reliability should be part of everyone's performance expectations.

Waste 7: We just can't improve our PdM work percentage

Problem:

Many organizations struggle to improve their condition monitoring work percentage despite sponsoring and funding a PdM team.

Potential Action A:

It is very common to have PdM resources pulled into unplanned work and even planned work that is not PdM. This occurs because supervisors see this crew as a resource pool. And again, the urgent always trumps the strategic. The manager needs to put in an approval process to pull PdM resources off of PdM. Further, have the PdM team report total hours worked on PdM each week. This metric needs to be charted for the reliability lead team.

Potential Action B:

Invest hours saved from PM kaizens and other efficiency projects into new PdM resources. For example: If you save 10,000 hours with a PM kaizen, you can move up to five persons to the PdM team for free. Five might not be the right answer, however; one or two might be better.

Potential Action C:

Start a PdM lite effort – as described earlier.

Potential Action D:

Require each planner to put at least one PdM follow-up work order on each week's schedule. This is to repair a known failure coming in the future. Too often the culture values time-based solutions.

Waste 8: PdM is just a hobby at my plant

Problem:

PdM is not seen as being on the same level as the proven PM process at our plant. We have a PdM team, but it is greatly under resourced and even discounted versus preferred invasive inspections.

Potential Action A:

Education is the best here. Decision-makers need to know the limitations of time-based maintenance and the benefits of condition-based maintenance. Thorough understanding of these practices only leaves you with one solution – PdM. Once you complete the education, quickly follow-up with a PM optimization process to cut out PMs that address a failure mode but are covered with PdM technologies. This is waste. Use the saved hours from all actions taken at the plant to further inflate the staffing of the PdM team. Far too often, maintenance managers only want to staff PdM teams after all PM and backlog actions are fully staffed. Leaders think they are being conservative with this approach, but in reality, they are accepting more risks and more failures due to ignorance and dedication to the status quo.

Potential Action B:

Have each member of the lead team sponsor a PdM technology. The expectation is for the leader to participate in a diagnostic route for at least 2 hours a month. In doing so, the mystery of the tool will fade, and confidence will build.

Waste 9: We keep starting and stopping our reliability efforts

Problem:

Top leaders believe the dogma that reliability programs require huge up-front investments with long term benefits. These investments can be stopped, and even paused, with minimal impact to the overall reliability journey.

Potential Action:

For this waste, education is key for top managers. Starting and stopping a reliability program that has a waste elimination focus makes no sense at all. If anything, the plant reliability leader should be asked how they can accelerate the process of eliminating waste. The leader should push back hard when asked to stop reliability efforts and deflect the request to a need to find financial savings. High level managers do not want to stop a good process; they want money. The leader may have to take on a waste challenge of 10% of the R&M budget, but that is not stopping your reliability effort. An "understood by all" waste focus deployment eliminates this problem.

Waste 10: Assigning craftsmen to open workorders

Background:

Out of convenience, it is common for a planner to create open work orders for employees that are assigned the same work assignment each week. Open work orders are work orders that for practical purposes never close. That is, each and every week 40 hours are charged per person to these jobs. Examples include pump repair, roller repair, and motor maintenance.

Problem 1:

Management has no control over the efficiency of the work. It is very common to find the workload of these individuals highly variable. They may have 60 hours of backlog to complete next week, and 10 the next. The planner assigns a craftsman to the area and leaves it up to the supervisor to pull him in or out of the job as work increases or decreases. This in itself is a problem, but let's look at reality. The supervisor is busy. Does he know the backlog of each person assigned to open work orders? No. Also, the person assigned to the open work order pushes back on the supervisor when they are asked to do other work. The "easy button" for the supervisor is to fill the new need for labor to overtime. I have found wrench time for open work orders to be the lowest in the plant for planned work. How big is the problem? One plant I was at had 75% of their crafts assigned to open work orders.

Problem 2:

Metrics. If a supervisor pulls a mechanic from an open work order job to an unplanned emergency job for two days this week, which work order do you think the mechanic's time gets charged to? You guessed it – it all goes to planned work. Now, repeat this for several crafts over several weeks; how accurate is your planned work percentage? unplanned work percentage? cost to maintain a given asset? The answer is: not very accurate.

Potential Action:

The simple solution is to have zero open work orders. The planner should evaluate the backlog for the area weekly and schedule manpower as necessary. However, I know this could be a major leap from where you are in current state. In such cases, assign a craftsman to an open work order for the lowest level work the area can see on a weekly basis. For example: A review of the backlog for the pump repair areas shows a backlog of 3 days, 4 days, 4 days, 3 days, 3 days, and 5 days. Create an open work order for 3 days per week. Work with the supervisor and employee to set triggers for increasing their scheduled time. By the way, over half the time the mechanic will figure out how to get the work done in the 3 days. This provides a boost in wrench time and in available resources. Either of these solutions fix your efficiency problem and your metrics problem. Lastly, all open work orders should be audited at least every 6 months to ensure this baseline of work has not drifted up or down. It is very common to set up open work orders for a current state that changes over the years and piles up waste unnoticed. It is best to make the backlog visual via a work board. This can be audited weekly by the planner before scheduling work.

You may be asking yourself, "Why don't planners and supervisors self-identify the waste in open work orders?" The answer is quite simple; if a planner has 20 craftsmen to plan for next week, the job gets a lot easier if 16 of them are assigned to open work orders. Now she needs to just plan for 4 persons. Plus, that's 80% planned work without accounting for the remaining persons. Management loves the high planned work metric. Part two of the answer is the benefit to the supervisor. Walking a mile in the shoes of a maintenance supervisor will give you great respect for the position. Demands for manpower are constant and everywhere; having a large pool of craftsmen "hidden" in planned open work orders solves a lot of problems for them. This is the "game" behind the scenes in maintenance once you question the accuracy of metrics. You may be saying that this is how the world works, and your plant is able to function in spite of management. However, all this work into which the supervisor and planner are moving resources is unapproved and has very low wrench time. The purpose of a planning meeting is for the owners to decide where they want to invest resources next week. In the example above, the planning meeting would only decide where the 4 persons not assigned to open workorders are assigned – that's 20% of the total resources. This is living a lie, and a system focused on KPIs and metrics is encouraging these bad practices.

Do you see why I don't trust KPIs? Think this is not happening in your plant? I suggest you find out through auditing and chalk circle observation. Correcting the waste in open work orders was huge in my plants not only from an efficiency standpoint, but also for our ability to make better decisions. For example: Do we need to replace a mechanic that is retiring? Do we need to replace an asset because we know the real cost of operation? Do we need to run at 20% overtime? Can we make 10% more product?

Waste 11: Our backlog is growing because we can't hire craftsmen

Problem:

I have heard this comment dozens of times. Unfortunately, these managers are not looking at waste; they are looking at the situation at current state performance, and the only way to solve it is to add manpower. I have performed a deep dive reliability assessment at 33 locations. I have never recommended adding headcount to solve a backlog problem – never. Solutions always came from wrench time improvement (kitting and staging), problem solving, PM kaizens, or increasing PdM.

For fun, let's say the maintenance manager did get approval to hire 3 persons to reduce backlog. How long would it take to hire, on-board, and train these three to become productive resources? 6 months? 9 months? Perhaps. So, we are waiting these 6 months, and then 2 persons go out on medical leave. Well, I guess we'll wait the 6+ months for both employees to come off medical. How many have played this delay game for years?

Potential Action:

How long would it take to select and train a kitter/stager role from existing headcount and improve wrench time by 10%? A month? This newfound efficiency nearly always dwarfs the impact of adding people to a poor efficiency system. Throwing manpower at a problem often appears to be the right answer, but in reality, it normally fails. I contend that a focus on waste via chalk circle observation will always be faster than hiring new resources. The kitter/stager is just one answer; there are countless other actions that will reveal themselves once you open your eyes to the waste at your plant.

A recent article I read stated that with the Baby Boomers now retiring, the industry will lose 50% of craft resources within 10 years. This can be scary to equipment-intense manufacturing businesses. At best, filling every craft opening will be very difficult in the future. You should also expect the salaries of these positions to rise due to supply and demand. Consequently, your plant needs to drive efficiency for survival as well as to improve performance. Leaders need to take action now.

Waste 12: New management procedures are impacting wrench time

Problem:

It is very common for management to make process changes without knowing the consequences of those actions on shop floor efficiency. Two examples:

1. A new requirement that lock/tag/verify procedures are not to be printed out for work packs before the day of work execution. This is to guarantee the most current version is being used. While good in intention, this requirement resulted in crews waiting up to an hour each morning to receive their final procedures. This was a real example.

2. A new form from safety titled "Approval to Proceed" was created by the safety department to ensure contractors had permission to begin work each day. The form was to be completed by the department supervisor (not the contractor). The supervisor was accountable for verifying all contractor precautions had been addressed. Again, good intentions, but this resulted in contractor crews waiting an average of 2 hours for this approval. Wait times ranged from 1 hour up to 4 hours. Why? The start time for the contractor coincided with the start time of the supervisor's normal crew. Once the supervisor got his crew started, he directed his attention to the contractor. This was a real example. The impact of this change was well over $1MM.

Potential Action:

Require chalk circle observations both before and after process changes to validate current state and impacts – including unanticipated consequences.

Waste 13: Vehicle availability – usage hours on fork truck

Problem:

Plants commonly evaluate their fleet of vehicles to control costs. Unfortunately, hours of operation is the most common metric used in this evaluation. For production, this is a pretty good tool. For maintenance, vehicles have peaks and valleys of operation; that is, they are not required on a steady basis. Consequently, it is easy for management to look at a fleet of vehicles assigned to maintenance and demand a decrease. Example: You have three fork trucks each utilized for one hour a day. It makes perfect sense to reduce them to one fork truck utilized three hours a day, right? The problem is that all three fork trucks are used the first hour of the shift to stage parts. Thus, the change to 1 fork truck will drive some crews' wrench time down by one to two hours a shift, three shifts a day, 30 days a month. The net negative effect can be 100 times the expected savings from reducing the fleet.

Potential Action:

Require chalk circle observations both before and after process changes to validate current state and impacts – including unanticipated consequences.

Waste 14: No morning meeting or standards of communication

Problem:

Getting all maintenance resources aligned on the priorities of the day is critical to success. It enables workers to realign their work and support each other in meeting goals. Further, keeping strategic goals and metrics in front of employees ensure all know what long term winning looks like. This gets them involved in problem solving plus gives their work meaning.

Potential Action:

A morning meeting is a very effective tool for alignment and communication. I prefer to have a standard meeting format that is detailed on a 4x8 foot white board. Begin with environmental, health, and safety; then cover KPIs for the last 24 hours; next, cover the priorities for the next 24 hours. Once a week, review strategic KPIs and project status. Keep the meeting under 15 minutes.

Waste 15: Not getting operators involved in reliability actions

Problem:

There is always more work to be done and too few maintenance resources – but don't forget about operations. At a minimum, they can be eyes and ears on the equipment reporting on changes in asset operation. Operators can also clean to inspect as part of their daily assignments.

Potential Action:

Implement TPM. Caution: Any actions by operators needs to be standardized and audited. It is very common for TPM efforts to start off great, but then drift away over time.

Waste 16: Not utilizing craft skills

Problem:

The current maintenance staff is short of the required number of planners, supervisors, PdM techs, maintenance engineers, and reliability engineers. We keep asking for approval to add headcount but are consistently denied.

Potential Action:

Assign your maintenance journeymen to these tasks and develop them over time. They may not be the perfect candidate today, but they are much better than leaving the position vacant. "Don't let perfect get in the way of good."

Waste 17: Program of the month

Problem:

"We start a new program every month." This is very common at plants. Somebody goes to a seminar or reads a book and now a new change cometh. Consequently, the change from last month lowers in status to make room for new plans. Now repeat this every month for 10 years, and you have a "program of the month" culture. This leads to confusion and a massive drift from standard work. It also breeds skepticism of organizational changes.

Potential Action:

Chalk circle observation of current state must become a requirement before new actions are taken. This is to ensure you understand the waste before you implement a solution. Secondly, an audit process of standard work must be created and utilized. Using chalk circle along with audits will eliminate the program of the month culture.

Waste 18: Focus.

Problem:

Management teams love KPIs. However, they forget the "K" in KPI stands for "Key." If your reliability lead team is tracking 64 KPIs, you are adding significant waste and confusion to your journey. (I actually did have one plant reporting on 64.) It is not reasonable to expect humans to know and improve upon dozens of metrics. At best, you can make glacial progress on several of these, but nothing meaningful.

Potential Action:

Warren Buffett, the investment mogul, has a great YouTube video on focus. I have details on his 5/25 rule in the Quick Reference Guide. In a nutshell, he states great success comes from the "focus on a few" mindset. Further, you are to ignore the trivial many. I believe a lead team should have 3-5 KPIs or initiatives they are driving at one time. Pick the right ones, and meaningful results will be your reward.

Waste 19: Adding PdM to PMs without eliminating the redundancy

Problem:

In the evolution of maintenance best practices, PMs came centuries before practical condition monitoring. Consequently, it is common to find PdM tasks simply added to existing PM functions. Example: In PM, every quarter we drain and clean the hydraulic oil from the system. In PdM, we sample the oil in the hydraulic system monthly for analysis. In this example, most likely the PM can be cancelled or at least dramatically modified. Risk-averse maintenance managers know of this overlap but are too cautious with change.

Potential Action:

Conduct a PM Optimization or kaizen to look for overlap of PMs with PdM. Default to the PdM technology. Maintenance managers need to learn to accept change risks. I suggest they work out an agreement with their manager to understand the risk/reward of moving to best practices.

Waste 20: Focus on new tools to buy without basics in place

Problem:

If you go to conferences, read trade magazines, or read LinkedIn posts on reliability, most of what you will see is new technology available to dramatically improve your results. Examples: A new smartphone software to track CMMS KPIs, AI, and IIoT. However, what you must understand is that all these technologies assume you are doing the basics well – basics such as planning and scheduling, work control, communication, kitting and staging, reliability engineering, problem solving, and precision maintenance. These technologies work great on top of the basics, but they are not a shortcut.

Potential Actions:

Ensure you assess the basic concepts of reliability at your plant before you purchase any new tools to accelerate your journey. I also recommend chalk circle observations of the problem you are seeking to fix. For most plants, technology is not your problem. Save your money.

Waste 21: Corporate reliability directives

Problem:

Every plant culture, business case, and equipment current state is unique. It is just wrong, ignorant, and arrogant for corporate to dictate reliability actions at the plant level without plant input and chalk circle observation. Corporate resources are talented and have great skills and passion to bring to your journey. However, this passion must be focused on the problem.

Example: A corporate mandate gets communicated to standardize your materials management process. Your plant has a very good materials management process already based on best practices and audited via chalk circle, but it is different than the corporate standard. Consequently, you are going to divert resources from your top 3 initiatives to support the mandate – waste.

Potential Action:

When you receive a corporate mandate, ask these resources to come to your plant for an evaluation of your current state and your reliability master plan. If you have a corporate resource, it is best to invite them to be part of your reliability lead team (a potentially remote participant). As such, they can be part of your chalk circle observations and plan creation.

Waste 22: Lack of precision maintenance

Problem:

Work is executed by the craft team, but they don't have the skills to do the work properly. I find this most common for lubrication, bearing installation, and shaft alignment. Every KPI and management decision assumes work is performed with precision. Example: If the mean time between failure (MTBF) of an asset is going down, management may start a funding request to replace the asset. In reality, lubrication practices have degraded, resulting in the increased failures.

Potential Actions:

Standard work for leaders to audit precision of work completed must become part of your reliability plan. Further, pay attention to RCA findings and look for trends with precision as a root cause.

Waste 23: Not making organizational changes

Problem:

On your reliability journey, you will have a small group of people who don't want to accept best practices. There are many reasons for this: they are comfortable with the way things are, they have excelled in the current state, and/or they are afraid they can't adapt and will fail.

Potential Action:

One-on-one coaching is a must in these situations. These individuals need to clearly know expectations. For the journeymen, keep selling your plans via examples. Spend time on the 90% who accept the hope for a better future you are providing. Don't focus on the 10% – let the 90% put pressure on them. Unfortunately, some members of your maintenance lead team will not be able to change, even with excessive coaching and training. You can't let them derail the future. It will be unusual if you do not have to make organizational changes as part of your journey. This can entail new assignments or employment termination. Obviously, the former is preferred.

Waste 24: Not understanding opportunity cost of poor reliability

Problem:

I could argue that the waste of lost opportunity due to poor factory reliability is the biggest financial loss your company is experiencing. Worse, you don't even know it. This is the waste of human resources focused on improvement and innovation that are distracted by reliability failures. Unfortunately, the impact of operational stability is nearly impossible to calculate. Plant leaders that have realized stability will always reference the "surprise" results of focusing resources on improvement rather than reacting to events.

Potential Action:

I recommend you note in the business case that opportunity cost is an "upside" to your estimates. A good leader will recognize the value, and it may tip the scale in their reliability decisions.

Waste 25: Not having active sponsors

Problem:

Knowledge of reliability best practices and availability of tools are rarely reasons cited for poor results. Lack of support for reliability in leadership's daily actions is the major killer of reliability efforts.

Potential Actions:

Start a reliability lead team consisting of top plant leadership positions. Include HR, IT, production, maintenance, engineering, and finance. Assign each member a specific accountability.

For example: the production manager leads the PdM team, while the HR manager leads the communication team. All sponsors must support reliability activities with their time. Leaders must attend planning meetings, OEE meetings, and create a "go and see" culture.

Leaders must incorporate reliability into their presentations – similar to safety, quality, and cost. Reliability must become how work is done, and not a program.

Waste 26: Adding discovery work time to PMs and other work orders

Problem:

This topic was introduced in the PM kaizen section earlier. It is so prevalent in plants that I chose to single it out to highlight its importance. I have also gone into more detail here.

If a planner and supervisor estimate a job will take 2 hours, most often they mutually agree to schedule the job for at least 4 hours. Their logic is one or more of the following.

1. 1 out of 10 times, we find discovery work and need the additional 2 hours. If there is no discovery work, we can just assign the team more work at that time. While this sounds reasonable, the assignment of new work rarely occurs – most often due to the supervisor being too busy or distracted. This is most prevalent in large, older plants. As PMs and repair work get executed over the years, more and more rare events creep in, further extending the scheduled duration.

2. Metrics. Most organizations use schedule compliance to measure planners and supervisors; consequently, there is pressure to schedule work for easy completion. Another measure is planned work percentage. If you plan a job for 6 hours, and it only takes 2 to execute, the remaining 4 hours can be used to execute the unplanned work, and the hours will be all recorded as planned.

3. Emergency work. Supervisors juggle unplanned work with planned work. This task is much easier if the planned work is padded with extra hours. This is further supported by leaders'

actions. How many times in a morning meeting has a leader recognized the maintenance team for completing extra jobs not on the schedule for the previous day? This positive reinforcement promotes putting waste in the schedule.

Potential Action A:

Auditing in the field as part of standard work will discourage these behaviors. Consider auditing jobs after the morning meeting each Tuesday.

Potential Action B:

Conduct a PM kaizen (optimization). I've detailed this process earlier in this book. This is a great way to get 50% of the opportunity captured in a very condensed time period.

Potential Action C:

Chalk circle observation of tasks needs to be part of every leader's standard work. Once you complete this task, don't forget to check how the metrics on the jobs you just observed were documented.

For example: You observe a crew for 8 hours, and they worked on planned work for 3 hours and unplanned work for 5 hours. How many hours are charged as planned work? There's a good chance it will be the full 8 hours rather than the accurate 3.

Waste 27: Thinking part-time reliability engineers are effective

This waste is particularly emotional with me. As the maintenance and engineering manager at a site, I had 10 engineers working for me. Each of them was assigned an area of the plant. They had full accountability as maintenance engineers, industrial engineers, capital project engineers, department technical assistants, supervisor fill-ins, and reliability engineers. I like the setup. We were a 24/7 operation, and the engineers were able to "reap what they sowed." If they did great reliability work, they got few calls on nights, weekends, and holidays. A corporate leader I'll call Vince argued with me for a full year to dedicate full-time reliability engineers to enable them to focus. I stood my ground. Observation revealed that in reality the engineers were being consumed by the drama of unplanned work. It was hit-and-miss at best for finding time to work on the strategic reliability issues of the area. After 12 months with limited strategic reliability results from this group, my lead team and I decided to run an experiment. We divided the plant into 8 areas and assigned each an engineer. We took the 2 remaining engineers and made them full-time reliability. They were not to attend daily production meetings, and no one at the plant was able to give them work except through me. The test was set up for 6 months, which we extended to 12 pretty quickly based on their results. Their expectations were simple; detail $2MM in cost avoidance. This was pretty easily accomplished.

Problem: The urgent issues of a plant will consume your strategic resources.

Potential Action:

Carve out at least one full-time person dedicated to the strategic reliability problems at your site. Have them use CMMS data along with other inputs to make meaningful progress on the problems that plague your plant year after year.

Example: If motor failures is the top issue for the last 10 years, have a person (and a part-time team) focus on this issue alone for 12 months.

Waste 28: Not closing out actions on RCAs (Root Cause Analyses)

Problem:

You all have read surveys conducted that state that 80% of corrective actions from root cause investigations are never implemented. This is not only dooming the organization to repeat the failure, but also poorly motivating the problem-solvers and overall organization.

Potential Action:

Implement an action tracker with a SPA (single point of accountability) and by-when dates. Review the status of these in the reliability lead team. There is readily available software that can help you manage this. Many programs will send the SPA an email when a project is coming due – and send an email to the supervisor when past due.

Waste 29: Not looking at connections between groups for waste

Problem:

It is common for supervisors to look for and eliminate waste within their work team; but, who is looking at connections between groups at your site?

For example: Driving efficiency within the mechanic fabrication shop. However, I have found a rich opportunity to look for waste between groups. Consider the connection when production transfers ownership of an asset to maintenance at the start of an outage. Often equipment is not in the correct position, has not been properly cleaned, or was not turned over at the agreed-upon time.

Potential Action:

During your chalk circle observations, keep a keen eye on all connections between groups and between shifts. This is a source of tremendous waste at plants. It is common to lose an hour of efficiency with each connection. What's great about these wastes is that the solution rarely requires capital expenses, and results can be rapid. Solutions are more related to standard work and expectations, which can be implemented in just days.

www.ingramcontent.com/pod-product-compliance
Lightning Source LLC
Chambersburg PA
CBHW071410210526
45465CB00001B/324